You are standing on a certain spot.
No matter which direction you walk,
you'll be going south. Where are you?
(See page 50)

You live in the Northern Hemisphere.
Your friend lives in the Southern Hemisphere.
When you are at school on a wintry day,
he is having a summer vacation.
And when winter comes to his part
of the world, it's summer in yours. Why?
(See page 32)

You are walking in the rain.
The rain is almost over. The sun comes out.
You look up and there is a rainbow. Why?
(See page 27)

to HOLLY

Joe Kaufman's

ABOUT THE BIG

SKY

ABOUT THE HIGH

HILLS

ABOUT THE RICH

EARTH

...AND THE DEEP

SEA

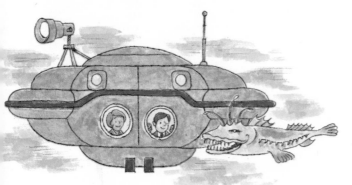

Written and Illustrated by JOE KAUFMAN

In consultation with CHRISTOPHER J. SCHUBERTH
Curator of Education, Illinois State Museum

Golden Press • New York
Western Publishing Company, Inc.
Racine, Wisconsin

Your Highnesses, I see the earth very much as ancient Egyptians did—as a vast room. The land and water are the floor. The sky, held up by columns, is the ceiling. The stars and moon are lanterns which light up each night.

So logical.

And so sound.

CONTENTS

A NOTE TO PARENTS: This book has been designed to help children learn about the world in which they live. Among many other things, children will discover how the sun causes Earth's weather, how mountains are formed, why Earth has seasons, how trees make their own food, why oceans are salty, what the water cycle is, how stalactites and stalagmites form, what galaxies are. The author-artist's approach, while completely accurate, is relaxed, humorous, and fun. The book is primarily for readers six to twelve years old, but its information has appeal for all ages.

What a nut! And he dares call himself a wise man?

Ridiculous!

Preposterous!

Your Most Royal Highnesses, I have the latest ideas on the subject. The earth is a spinning ball. People at the top walk about as we do. Those at the bottom walk upside down as flies do on a ceiling. On the sides they sort of stick out and walk about as flies do on a wall.

6

The **EARLIEST IDEAS** people had about the earth, sun, moon, and stars are lost because there was no written language. When people died, their ideas were gone, and we can only try to guess what those ideas were. Early humans had no way to travel to places far from home, so they probably thought the world was much like the place where they lived. To a jungle tribe, the earth must have seemed to be a hot, endless jungle. To hunters on the plains, the flat and grassy earth must have seemed to stretch on forever.

Then, as civilization began and written language developed, people began to write down stories to explain the world around them. Some Egyptians thought that the earth was a huge god, lying down. Trees and plants grew from his body. Arched above him was an equally huge goddess. She was the sky, the stars her jewelled robes. Babylonians im-

agined that the earth was inside a hollow mountain.

As time went on, some people began to make observations of the real world. Aristotle knew that the world was round—he had seen its round shadow on the moon. Copernicus showed that the sun was the center of the solar system, and the earth and the other planets revolved around it. But it was a long time before most people could understand and accept the ideas that came from these observations.

It is only in the last 100 years that scientists have really begun to explore the earth and the universe. Today, scientists dig deep holes into the earth's crust. They dive to the bottom of the ocean in laboratory-submarines. They blast satellites into space to study the earth's atmosphere, travel to the moon, and land instruments on planets. Some of the things they have discovered about our world and its surroundings are in this book.

When some people look up at the sky at night, they see nothing but the stars scattered all around.

A STAR is a sun so far away that it seems tiny.

A CONSTELLATION is a group of brighter stars.

A GALAXY is made up of many billions of stars.

THE UNIVERSE is the many billions of galaxies.

Look up at the STARS on a clear night when the moon isn't visible — you can see almost 3,000 of them without the help of a telescope.

Early people thought that certain groups of stars formed outlines of people or animals in the sky, and they gave them names. We call these star-groups "constellations."

People north of the equator see mostly the stars of the Northern Sky; people south of the equator see mostly the stars of the Southern Sky. People at the equator see almost all the stars.

The stars seem to move slowly from east to west because the earth is turning slowly from west to east. It turns about its axis — an imaginary straight line which runs through its center from North Pole to South Pole. Polaris, the North Star, is almost directly over the North Pole end of the

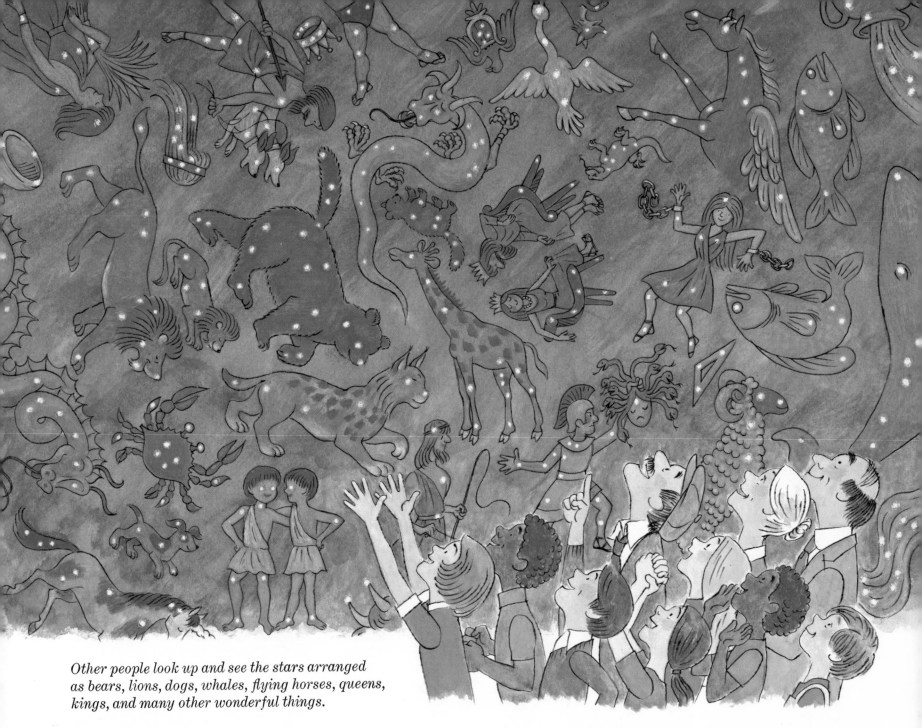

Other people look up and see the stars arranged as bears, lions, dogs, whales, flying horses, queens, kings, and many other wonderful things.

axis, so it seems to stand still, with the other stars moving around it.

Across the sky is a cloudy band called the Milky Way. It is what we on Earth can see of our very own galaxy. With binoculars, we can see that it is made up of separate stars, too many to count. The stars in our galaxy (our sun is one of them) form a huge pancake shape, with a bulge in the middle. Scientists estimate that our galaxy is 100,000 light years in diameter. A light year is the distance light travels in one year. In just one second, light travels 186,000 miles, so you can see how vast a distance even one light year is. Now try to imagine 100,000 light years!

With powerful telescopes we can see many of the other galaxies in the universe. The nearest one, Andromeda, is two billion light years away. A rock-et going at the speed of light (about 25,000 times faster than any rocket we have today) would take two billion years to get to Andromeda from Earth.

We don't know how far the universe stretches, but we do know that it has billions of galaxies in it.

The ZODIAC is a group of twelve constellations. They form a path circling the sky. The sun, moon, and planets seem to travel through this path.

Scorpio Libra Virgo Leo Cancer Gemini Taurus Aries Pisces Aquarius Capricorn Sagittarius

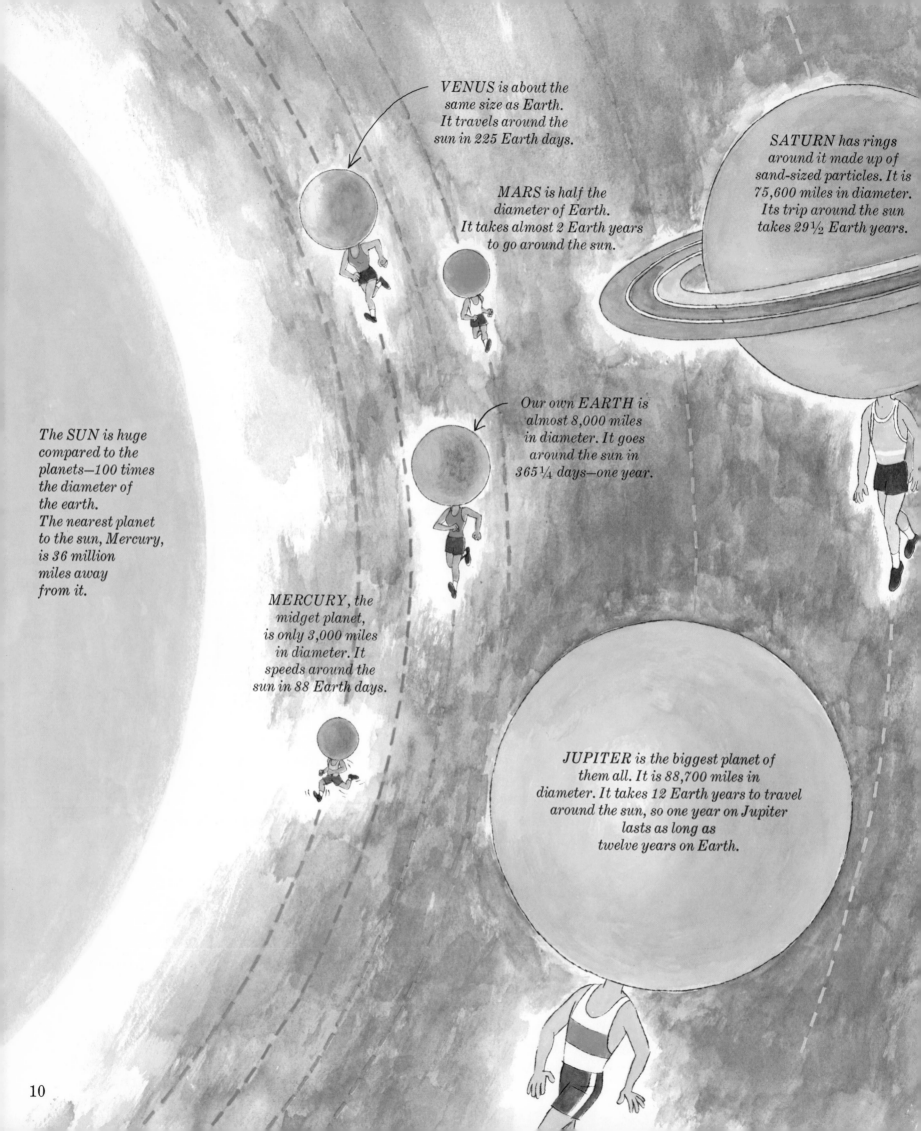

VENUS is about the same size as Earth. It travels around the sun in 225 Earth days.

MARS is half the diameter of Earth. It takes almost 2 Earth years to go around the sun.

SATURN has rings around it made up of sand-sized particles. It is 75,600 miles in diameter. Its trip around the sun takes 29½ Earth years.

The SUN is huge compared to the planets—100 times the diameter of the earth. The nearest planet to the sun, Mercury, is 36 million miles away from it.

Our own EARTH is almost 8,000 miles in diameter. It goes around the sun in 365¼ days—one year.

MERCURY, the midget planet, is only 3,000 miles in diameter. It speeds around the sun in 88 Earth days.

JUPITER is the biggest planet of them all. It is 88,700 miles in diameter. It takes 12 Earth years to travel around the sun, so one year on Jupiter lasts as long as twelve years on Earth.

10

URANUS is 29,500 miles in diameter. It orbits the sun in 84 Earth years. It, too, has rings.

NEPTUNE is almost as big as Uranus. It circles the sun in 165 Earth years.

PLUTO, the last planet to be discovered, is about 3,600 miles in diameter. Farthest from the sun, Pluto takes the longest time to travel around it—248 Earth years.

The nine PLANETS, their thirty-two moons, and the sun in the center are our solar system. The planets move in their own paths around the sun, and the moons move in their own paths around their planets. All these movements are controlled by gravity (see page 18).

The planets Mercury, Venus, Earth, and Mars are nearest to the sun and have solid surfaces. They are small compared to Jupiter, Saturn, Uranus, and Neptune, which are made mostly of gases. Pluto, another small planet with a solid surface, is farthest from the sun.

Although planets look like stars to us on Earth, they have no light of their own. They reflect the light of the sun shining on them. Stars, each one a sun like ours, shine with their own light.

Mercury, the smallest planet, is nearest to the sun. You can see it at dawn where the sun is about to rise and in the evening where the sun has set.

Venus has a thick cloud cover which reflects so much sunlight that Venus sometimes looks brighter than the brightest star. Its surface temperature is well above the boiling point of water.

Earth, that special planet we live on, is still the only planet we know of with life on it.

Mars looks reddish to us on Earth, and the color TV pictures taken by a camera we sent to its surface show us a view of reddish rocks on a red, deserted plain. Mars has polar ice caps which seem to grow and shrink as the seasons change.

Jupiter is the biggest planet of all. It spins around on its axis every 10 hours.

Saturn, the second largest planet, is known for its rings, made of sand-sized particles of ice.

Uranus is thought to have rings like Saturn's, and Neptune is hidden by clouds.

Pluto was the last planet to be discovered. It was found in 1930 and is the most distant planet in our solar system. Because it is farthest from us, it is least visible and we know least about it.

Each planet's year is the time it takes that planet to circle the sun . . .

so a four-year-old on MERCURY would be 1 year old on EARTH,

and a one-year-old on PLUTO would be 248 years old on EARTH.

These flame-like masses of gas that shoot out from the sun's surface are called prominences.

The temperature of the glowing surface of the sun is 10,000 degrees Fahrenheit (5,500 degrees Celsius). But at the center of the sun, scientists think the heat reaches 27 million °F (15 million °C).

Sometimes dark sun spots that may be storms appear on the sun's surface.

the earth ⟶

the 8 other planets

12 the 32 moons

All the planets and the moons around them add up to less than 1 per cent of the solar system's matter. All the rest of the matter, more than 99 per cent, is in the sun.

The SUN is just an ordinary star in space. It seems unlike a star to us on Earth because the sun is so near to us and the other stars are so far away. The sun is a mass of fiery hot gases, almost all hydrogen and helium, that is hottest in the center. This center of the sun, the source of all its heat and light, is an enormous nuclear energy furnace. It has been releasing nuclear energy for billions of years and will continue releasing it for many more billions of years. The heat in the sun's center is measured in millions of degrees and on the cooler surface in thousands of degrees.

Of all the enormous quantities of heat and light that the sun constantly sends out, only a tiny part reaches the earth. The rest is scattered in all directions. But the tiny part that does reach us is all-important, for without it no life would exist on the earth.

The light from the sun is so intensely brilliant that your eyes can be permanently damaged if you look directly at it. No one should ever look directly at the sun. Special equipment has been invented so that scientists can study the sun without hurting their eyes. Photographs are often used to study sunspots.

The sun turns around on its axis just as the earth does. The earth makes a complete turn once a day; the sun turns once every 25 days. The earth travels continually around the sun and takes a year to make the trip. The sun travels continually around the center of our galaxy and takes 200 million years to make the trip.

The sun's mass (the amount of matter in it) is almost all the mass in the solar system and it has a tremendous gravity pull. This gravity pull holds all the planets in their paths around the sun. Gravity on the sun is so strong that someone weighing 100 pounds on Earth would weigh 3,000 pounds (a ton and a half) on the sun.

If you could be far enough out in space to look back, the sun would be just another star.

The 93 million mile trip to the sun would take about 5 months, but no one would make such a trip. Any vehicle coming within 50 million miles of the sun would burn up.

Why do the sun and moon look the same size when the sun is 400 times as wide?

Because the sun is 400 times farther from Earth than the moon is.

If the sun were a hollow ball, it would take more than a million Earth-sized planets to fill it up.

Scientists study the sun (really a star) to understand more distant stars. This astronomer's telescope has a dense filter to protect his eyes from the sun.

Every month the moon goes through these phases. First it is all dark (new moon). Each night we see more and more

of it until it is a full moon. Then we see less and less of it each night until it is back to the all-dark, new-moon phase.

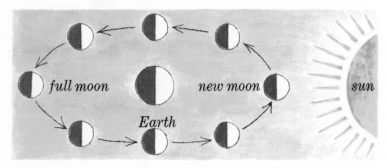

The moon's phases depend on how much sunlight the moon reflects to Earth and the positions of sun, moon, and Earth.

We always see the same side of the moon as it travels around the earth. Astronomers say this happens because there's a bulge on the moon, and Earth's gravity keeps that bulge turned toward us.

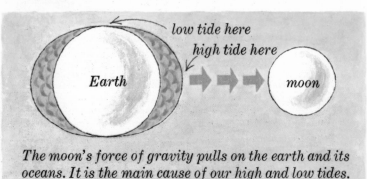

The moon's force of gravity pulls on the earth and its oceans. It is the main cause of our high and low tides.

The **MOON** is about one-fourth the size of the earth. No other planet has a moon so near its own size. If you were on Mars looking toward Earth with a telescope, Earth and the moon would look like two planets, one smaller than the other.

There are many theories of how the moon was formed, none that are completely accepted by all astronomers. One idea is that billions of years ago a big chunk of the earth broke away and became the moon. Another theory is that the moon was formed at the same time the earth was formed and of the same material. Still another is that the moon was formed in a distant part of the solar system and then, at a later time, passed near the earth and was captured and held by Earth's gravity.

We here on Earth see the moon mostly because it reflects sunlight. When you are looking at a full moon on a clear night, you see brighter and darker areas. The brighter areas you see are areas that reflect more sunlight. The darker parts reflect less sunlight. Light reflected from Earth also illuminates the moon.

Liebnitz Mts.— the moon's highest 26,000 ft. high

Mt. Everest— Earth's highest 29,028 ft. high

Looking through a telescope you see that the moon is covered with craters. There are millions of them of all sizes, some as big as 150 miles across. They may be scars from collisions with meteors.

The moon goes around the earth in about one month and turns on its axis in the same length of time, always keeping the same side toward us. It also goes around the sun with the earth once a year.

In 1969, our astronauts started a series of personal visits to the moon. They walked and drove around on the surface, set up scientific equipment there, and sent complete TV coverage back to Earth. They also brought back photographs, moon rocks, and soil. Now we have on-the-spot knowledge about a part of the universe other than Earth, and know just what it is like to feel a gravity force that is different from our own. On Earth, our astronauts, wearing all their equipment, each weighed 300 pounds. But on the moon, each astronaut weighed only 50 pounds.

The earth, almost hidden by clouds, as seen from the moon.

plastic plants and flowers, of course

TIPS and FACTS for moon settlers:
1. Bring air, water, and food—there's none here.
2. Each month there are 2 weeks of hot daylight (up to 212°F, 100°C), then 2 weeks of frigid darkness (down to 240°F below zero, −151°C).
3. Avoid the big craters when you drive around in your lunar rover.
4. Everything is silent on the moon—there's no air to carry sound. To say hello to a friend, use a walkie-talkie.
5. Remember the folks back on Earth—238,857 miles away.

This moon auto is called a lunar rover.

15

Earth is a small planet in our family of planets. Earth has 1 moon; Pluto also has 1. Mercury and Venus have none. Mars has 2; Uranus, 5; Saturn, 22; Neptune, 3; and Jupiter, 17.

This is the earth as we know it, with its land masses and bodies of water. This familiar view shows only Earth's thin outer surface of soil and water, covering the crust.

Earth weighs 13,200,000,000,000,000,000,000,000 pounds.

The EARTH is the only planet we know of with just the right environment for people, animals, birds, fish, flowers — all our living things — to exist. There is soil, filled with nourishment, lying on top of the earth's hard rock crust. There is the atmosphere, rich in life-sustaining gases, surrounding our globe. There is water, essential to all living things, covering most of the surface of the earth. The temperature, too, is just right; if the earth were farther from the sun it would be too cold for our kind of life to be possible, and if the earth were closer to the sun it would be too hot.

The earth is round. We no longer need scientific

Poke a ruler through the earth. The diameter will measure about 8,000 miles.

Now wrap a tape measure around the earth. It will measure 25,000 miles.

The earth spins around on its axis like a top, but it spins slowly. It turns completely around once every 24 hours.

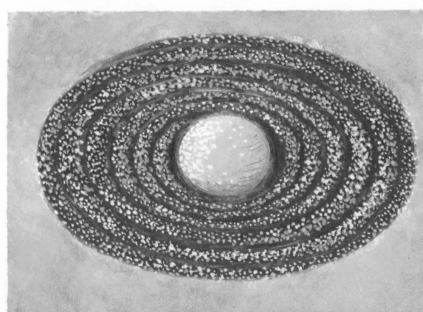

The earth and all the planets may have begun as whirling masses of dust and gas. Gravity began to pull each mass together into a ball which grew larger and still larger.

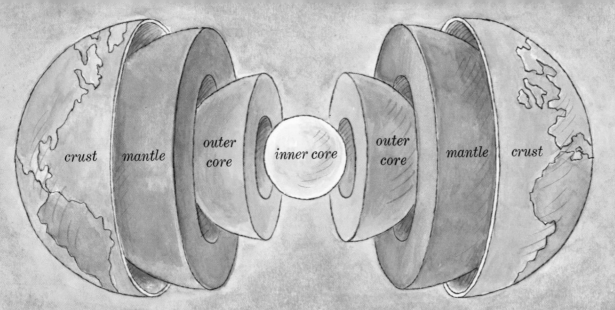

The crust, under the soil and water, is rock, 5 to 20 miles thick. The mantle is hot, heavy, bendable rock, 1,800 miles thick. The outer core is molten iron, 1,300 miles thick. The inner core is a scorching iron ball, 1,600 miles across.

An egg has layers, too. It has the hard shell, the egg white, and the core, its yellow yolk.

explanations to prove it. Now we can look at the photographs taken by the astronauts from outer space and actually see our round Earth.

The movements of the earth and moon give us our ideas about time. The amount of time it takes Earth to spin on its axis is 24 hours, which we call a day. The amount of time it takes the moon to circle the earth, we call a month. The amount of time it takes the earth to move around the sun, we call a year.

The North and South Poles (the geographic poles) are the places where the north and south ends of Earth's axis are located. The earth also has a north magnetic pole and a south magnetic pole (about a thousand miles from the geographic poles). These magnetic poles form a magnet that causes the needle of a compass to point north. Ships at sea, planes flying above, boy scouts and girl scouts, all rely on compasses and the magnetic poles to help them find their way.

The sun lights up half of the earth at a time. In the lighted part, it is daytime. In the dark part, it is night. Because Earth keeps turning on its axis, half of it is always moving from night into day, while the other half is always moving from day into night.

The new earth was a ball with a hard rock skin. Later, lava and gas burst out through the skin. The gas formed air and clouds. For millions of years rain fell.

Rain collected in low places, forming the oceans. The higher lands stayed dry. As the earth kept changing, mountains, plateaus, and plains formed on the lands.

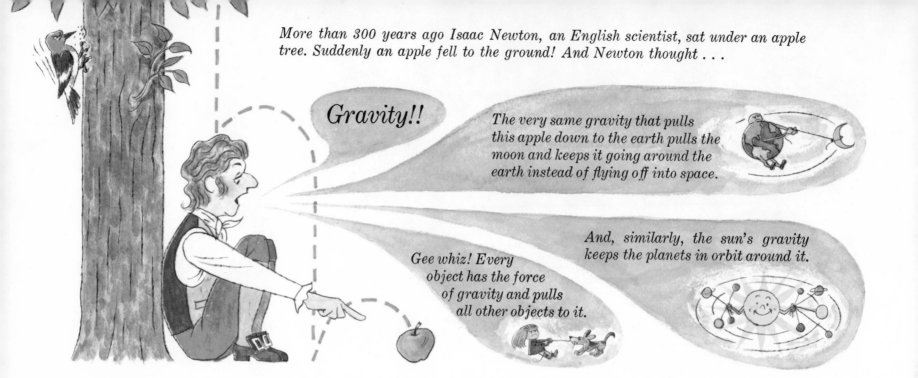

More than 300 years ago Isaac Newton, an English scientist, sat under an apple tree. Suddenly an apple fell to the ground! And Newton thought . . .

Gravity!!

The very same gravity that pulls this apple down to the earth pulls the moon and keeps it going around the earth instead of flying off into space.

And, similarly, the sun's gravity keeps the planets in orbit around it.

Gee whiz! Every object has the force of gravity and pulls all other objects to it.

GRAVITY is the invisible force that makes things fall when you drop them. When you drop a ball, it falls because the gravitational attraction of the earth pulls the ball toward the center of the earth. If holes were drilled straight down on opposite sides of the earth and a stone was dropped into each hole, each stone would fall to the bottom of its hole. This would happen because the force of gravity at the earth's center would pull both stones toward the center.

Every object in the universe, including the sun, moon, and planets, attracts every other object, and the greater the mass of the object, the greater the force with which it attracts other objects.

Did you know that you attract other objects? You do, but it's hard to tell that you do. For instance, if you hold a ball out at arm's length and let go of it, your own force of gravity pulls the falling ball in toward your body. But the earth's pull is so much greater than yours that the ball actually drops almost straight down to the ground.

Another thing about gravity is that the farther two objects are from each other, the weaker their pull on each other is. The sun, because it is bigger than the earth, has a much stronger gravity pull. And yet people standing on the earth do not go fly-

The earth's gravity pulls everything and everyone "down," or toward the center of the earth. What we call our weight is the force with which gravity pulls us.

The moon, smaller than Earth, has less gravity pull. There an astronaut weighs about 50 pounds with his pack and can hop like a rabbit.

But if an astronaut lands on Jupiter, our largest planet, he'll have to crawl. Jupiter's gravity pull is so strong he'll weigh about 700 pounds with pack.

ing off toward the sun. The earth's pull on us is stronger than the sun's because the earth is so much nearer; we're standing on it.

To reach the moon, the astronauts in the Apollo spaceship first had to escape from Earth's gravity. They did this by going at the tremendous speed of 25 thousand miles an hour. (A fast jet plane goes about 1½ thousand miles an hour.) When the spaceship reached this great speed, it broke free of Earth's gravity and then coasted through space toward the moon until it was captured by the moon's gravity and held in an orbit around the moon, just as the planets are held in orbits around the sun.

If gravity were ever turned off, there would be no force holding us down. We would be weightless and we'd float like astronauts in space. What fun!

FOOD from the SUN

The energy in our bodies comes from green plants. The plants' energy comes from the sun, and plants need sun to grow well.

Our food is either vegetable or animal. The vegetable part of what we eat comes, of course, from plants. The animal part of what we eat, whether it is meat, fish, milk, cheese, or eggs, comes indirectly from those sun-energized plants.

This is the sun that sends energy to plants.

This is the plant that stores the sun's energy.

This is the cow that eats the plant.

POWER FROM THE SUN

LIGHT POWER, HEAT POWER AND POWER TO GO! GO! GO!

Millions of years ago, there were vast forests of palm trees and other tropical plants. Dinosaurs roamed through them. As the trees and plants died, they sank down into deep layers of sand and mud.

Weight from the mud and sand that buried the vegetation with the sun's energy in it, hardened the vegetation into coal. Now we mine the coal and burn it to release the energy as heat. We also change the heat to electricity.

HOW THE SUN HELPED PUT NEIL ON THE MOON

Neil Armstrong, Buzz Aldrin, Mike Collins— full of energy from sun-powered food

The 2½ minute first stage of the rocket used 2,000 tons of kerosene fuel and liquid oxygen. Kerosene comes from good old sun-energized oil.

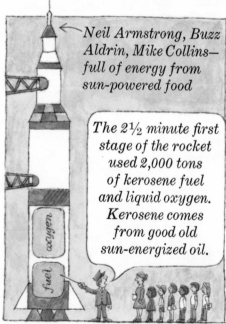

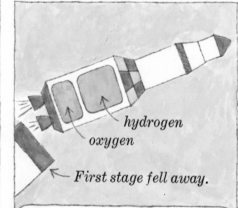

hydrogen
oxygen

← *First stage fell away.*

The rocket's second stage used liquid hydrogen fuel and liquid oxygen. Hydrogen is often made from methane or coal gas, both containing energy from the sun.

sun-powered fuel again!

liquid oxygen
liquid hydrogen

Third stage propelled astronauts directly toward the moon.

Second stage fell away.

This is the girl

that milks   *the cow.*

This is the cheese that's made from the milk. (It's still full of the sun's energy.)

So the energy we need to play, study, and work comes from our food. And all that energy in food came to it from the sun.

Similarly, oil and gas were formed from decayed prehistoric animals and plants in swamps and seas. Today we're using up that stored sun's energy very quickly.

OIL!!   OIL!!

So the sunshine that fell on Earth so long ago is now used for . . .

light . . .

heat . . .

going places.

landing module (LM) here

command module

It turned around and docked with lunar module, then third stage fell away.

Neil and Buzz climbed into LM and set it down on the moon.

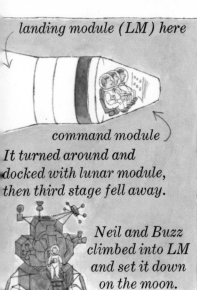

. . . one giant step for mankind

So, with a little help from the sun, our Neil Armstrong became the first person to walk on the moon.

ENERGY from the SUN, also called solar energy, makes all life on Earth possible. The sun's energy is used and stored by plants; animals eat plants; and people in turn eat both plants and animals.

The sun's energy determines the weather. We get rain or snow when the sun's heat turns ocean water into vapor which rises and condenses into clouds, then falls back to Earth. Winds start when sun-heated air rises and moves toward cooler places, to be replaced by cooler air which flows in underneath it. The spinning earth adds more motion to this air, which we call wind.

The sun provides us with light during the day, and power to make light at night. We also use the sun's energy stored in coal, oil, and gas to heat our homes.

In one daylight hour, enough of the sun's energy reaches the earth to supply all of the power to run vehicles and machines, to build and keep up our homes, heat and light our schools, run our stores and factories, for 1½ years. Most of this energy goes to waste because we don't yet know how to capture it.

Scientists are at work designing solar furnaces, harnessing wind power, making huge mirrors to collect the sun's rays, and trying to find ways to use the heat stored in the oceans.

The number of homes with water supplies heated by solar energy is increasing. Sunlight heats up metal panels on top of these houses, the panels then heat water in pipes behind them. The hot water is stored in a tank until you turn on the faucet.

Today, research into more and better ways of directly using the sun's almost unlimited energy is becoming more and more important as Earth's supply of gas, oil, and coal dwindles.

AIR
(Atmosphere)

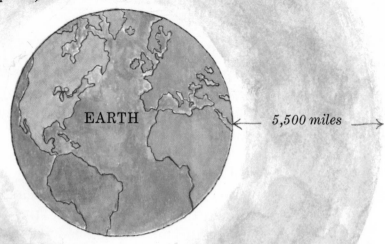
EARTH

← 5,500 miles →

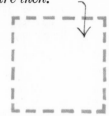

The AIR, or atmosphere, is a thick layer of gases surrounding the earth. Without the protection of the atmosphere, the earth would broil by day and freeze at night. In the daytime the atmosphere screens out the sun's most harmful rays, and at night it keeps most of the earth's heat from escaping into space.

The air is 78% nitrogen and 21% oxygen. The other 1% is made up of small amounts of other gases and pollen and dust. The amount of dust and other things in the air that make the air dirty (pollutants) varies from place to place. In smoky big cities there are days when the air is so dirty it's unhealthy for us to breathe it.

You cannot see air — it's invisible. Smoky, polluted air seems visible, but it's only the pollutants in the air that you see.

The atmosphere has four layers. The layer nearest the earth, called the troposphere, ranges from 5 to 11 miles in thickness. All of our weather takes place there. The next 40 miles, where there is no weather, is called the stratosphere. Jet pilots fly up there to avoid storms. The next layer, the ionosphere, extends out from the stratosphere to a distance of about 650 miles. The air there is thinner because the air particles are farther apart. The outermost layer is called the exosphere. It reaches out to 5,500 miles from the earth. The air there is very, very thin because the air particles are even more widely separated. Beyond the exosphere lies outer space, where there are almost no air particles.

Air has weight and presses down on the earth. At sea level there is about 15 pounds of pressure on every square inch of each person, place, and thing. You can't feel the weight of the air pressing on your body because it's balanced by the same amount of pressure inside your body.

The higher you climb, the less air pressure there is. The highest mountain peak has only 5 pounds of pressure per square inch.

If there were no air . . .

. . . there would be meteors falling all around us. Air's friction burns most of them up before they land.

. . . there would be no food. Air is necessary to make things grow.

. . . there would be no rainbows, as they are caused by sunlight passing through air's moisture.

People and animals need and breathe in oxygen, which plants give off . . .

. . . plants need and absorb carbon dioxide, that people and animals breathe out.

We cannot see air but we know it's there. Air keeps the plane, balloons, and all the rest flying and soaring. It turns the pinwheel and the windmill and blows the boat through the water.

This happy weatherman is using a very simple weather map. The dash lines show temperatures across the country.

The WEATHER is the daily condition of the troposphere in a particular place. What we call weather is air temperature, air pressure, how fast the air is moving, and how much moisture is in it. Weather concerns everybody — we try to dress to suit the weather, we build our homes to protect ourselves from it, and our food crops depend on it.

Temperature varies according to where you are. On the same day the temperature might be over 100°F (38°C) at the equator and 100°F below zero (−73°C) at the South Pole. It might be 70°F (21°C) at the foot of a mountain and 30°F (−1°C) at the top. The temperature of any area differs according to the season of the year. It's warmer in summer than it is in winter, even at the poles.

Although air pressure is about 15 pounds per square inch at sea level, it does vary somewhat as the weather changes. We say the air pressure is "high" when air particles are close together, making the air heavy. This is usually true when air is cool. We say the air pressure is "low" when the air particles are farther apart, and the air is lighter. This is usually true when air is warm.

Wind is air in motion and the amount of wind is an important part of weather.

Moisture in the atmosphere comes from water that evaporates, mostly from the oceans. Warm air holds more water vapor than cold air. We measure the amount of water vapor in the air and call it the "humidity."

A HOT, HUMID DAY is uncomfortable because the air is so full of moisture that your perspiration can't evaporate and cool you. You feel hot and sticky.

A SNOWY DAY happens when water vapor in the clouds freezes into snowflakes without first turning into water. Then the snow falls.

A RAINY DAY occurs when tiny drops of water in the clouds gather around specks of dust until they become larger drops heavy enough to fall as rain.

We have a FOGGY DAY when clouds lie at or near the earth's surface. In a fog cars go slowly and ships blow foghorns to warn other ships.

A PERFECT SUNNY DAY means no rain clouds, not too hot or cold, and humidity just right.

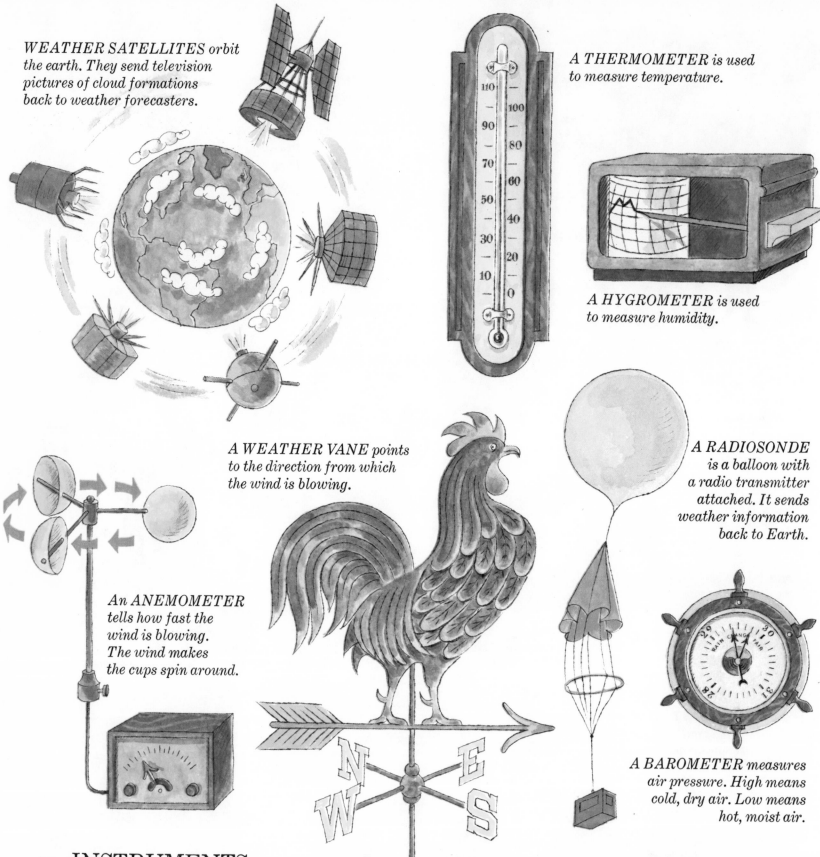

WEATHER SATELLITES orbit the earth. They send television pictures of cloud formations back to weather forecasters.

A THERMOMETER is used to measure temperature.

A HYGROMETER is used to measure humidity.

A WEATHER VANE points to the direction from which the wind is blowing.

A RADIOSONDE is a balloon with a radio transmitter attached. It sends weather information back to Earth.

An ANEMOMETER tells how fast the wind is blowing. The wind makes the cups spin around.

A BAROMETER measures air pressure. High means cold, dry air. Low means hot, moist air.

The INSTRUMENTS that weather forecasters use help them to make accurate predictions. Before the invention of weather instruments, farmers, sailors and others who worked outdoors had to guess what the weather would be like and some of them learned to make pretty good guesses.

Now, many different kinds of scientific equipment to measure temperature, direction of wind, wind speed, air pressure, moisture in the air, and amount of rainfall have been invented. Scientists called meteorologists, who specialize in studying weather, are trained to use these instruments and the information that weather satellites and radar transmitters send to them.

When forecasters make up their reports and maps, they use information from their own instruments and also information from the instruments of observers in other areas.

CIRRUS CLOUDS
(curly clouds)
no rain likely

A pilot can avoid rain by
flying over the rain clouds.

ALTOCUMULUS CLOUDS
(high, heaped-up clouds)
occasional rain possible

CIRRUS means CURLY
CUMULUS means HEAPED-UP
ALTO means HIGH
STRATUS means SPREAD-OUT
NIMBUS means RAINSTORMY
(Latin) . . . . . (English)

ALTOSTRATUS CLOUDS
(high, spread-out clouds)
rain or snow possible

STRATOCUMULUS CLOUDS
(spread-out, heaped-up clouds)
drizzle or snow flurries may fall

CUMULUS CLOUDS
(heaped-up clouds)
"fair-weather" clouds—no rain likely

NIMBOSTRATUS CLOUDS
(rainstormy, spread-out clouds)
cover the entire sky
steady rain or snow

CLOUDS are formed when the sun evaporates water (turns it into vapor) from oceans, lakes, and rivers. Trees and moist earth are also sources of water vapor. The vapor rises and becomes cooler. When it is cool enough it condenses — it changes into tiny droplets of water. Each droplet forms around a speck of dust in the air. Millions and millions of these droplets, gathered in a big mass, form a cloud.

In a cloud, water droplets grow by joining each other. When they become big enough, they fall as raindrops (a raindrop is at least 1/125 of an inch across). When it is cold enough, the droplets become snowflakes and it snows.

Sometimes, when a shower is ending and the sun comes out, you see a rainbow. The raindrops in the air act as prisms, separating the sun's rays into different colors.

Hot air begins to rise from warmer surfaces of the earth.

Hot, rising air, bearing lots of moisture, creates a cloud.

The electrically-charged cloud grows to thunderhead shape.

Rain falls, then electricity causes a flash of lightning.

LIGHTNING

LIGHTNING is a huge electric spark that flashes from a thundercloud. Lightning happens when strong winds inside a thundercloud whip its water drops so that they bang and crash against each other. All this banging and crashing creates electrical charges inside the cloud.

Some of the electrical charges are positive and some are negative. Great masses of negative charges are forced up to the top of the cloud. Some negative charges remain in the lower parts of the cloud along with some positive charges. Other positive charges are carried down to the ground by raindrops.

When great numbers of negative charges have gathered in the upper part of the cloud they suddenly shoot out in a spark to another part of the cloud, or to another cloud, or down to the ground—anywhere there are great numbers of positive charges. The giant electrical spark makes the air around it glow, and that is how we see lightning.

Forked lightning is the zigzag kind that branches out like a tree. Sheet lightning lights up most of the sky. Bead lightning looks like a string of beads.

Buildings are protected from lightning by lightning rods. People protect themselves by knowing what to do and what not to do in a lightning storm.

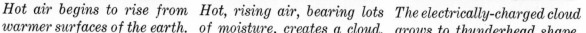

The ancient Greeks and Romans believed that the god Zeus (called Jupiter by the Romans) threw down lightning bolts.

THUNDER

THUNDER is the sound we hear after a flash of lightning. It is caused when the spark of lightning streaks out and heats the air around it. The heat is tremendous and the air expands violently, sending the particles of hot air flying out in all directions. They bang into the cool surrounding air, making great sound waves that we hear as the sound of thunder.

Lightning and thunder happen at the same time, but the light of lightning reaches our eyes much faster than the sound of thunder reaches our ears. The speed of light is about 186,000 miles a second, but sound travels only about 1/5 of a mile a second. So we see lightning that's a mile away almost immediately, but it takes 5 seconds for the noise of the thunder to reach us.

Sometimes we hear thunder when we haven't seen any lightning. Actually there was lightning, but it was too distant to see or hidden from us by trees or buildings.

Nearby lightning causes loud, sharp thunder. Rolling thunder means that the lightning is far away. If lightning is more than 15 miles away we can't hear its thunder at all.

Some North American Indians believed in a bird-shaped god of thunder called Thunderbird, whose flapping wings produced rolling thunder.

If you are 1 mile from the lightning, it will take 5 seconds for the thunder to reach you.

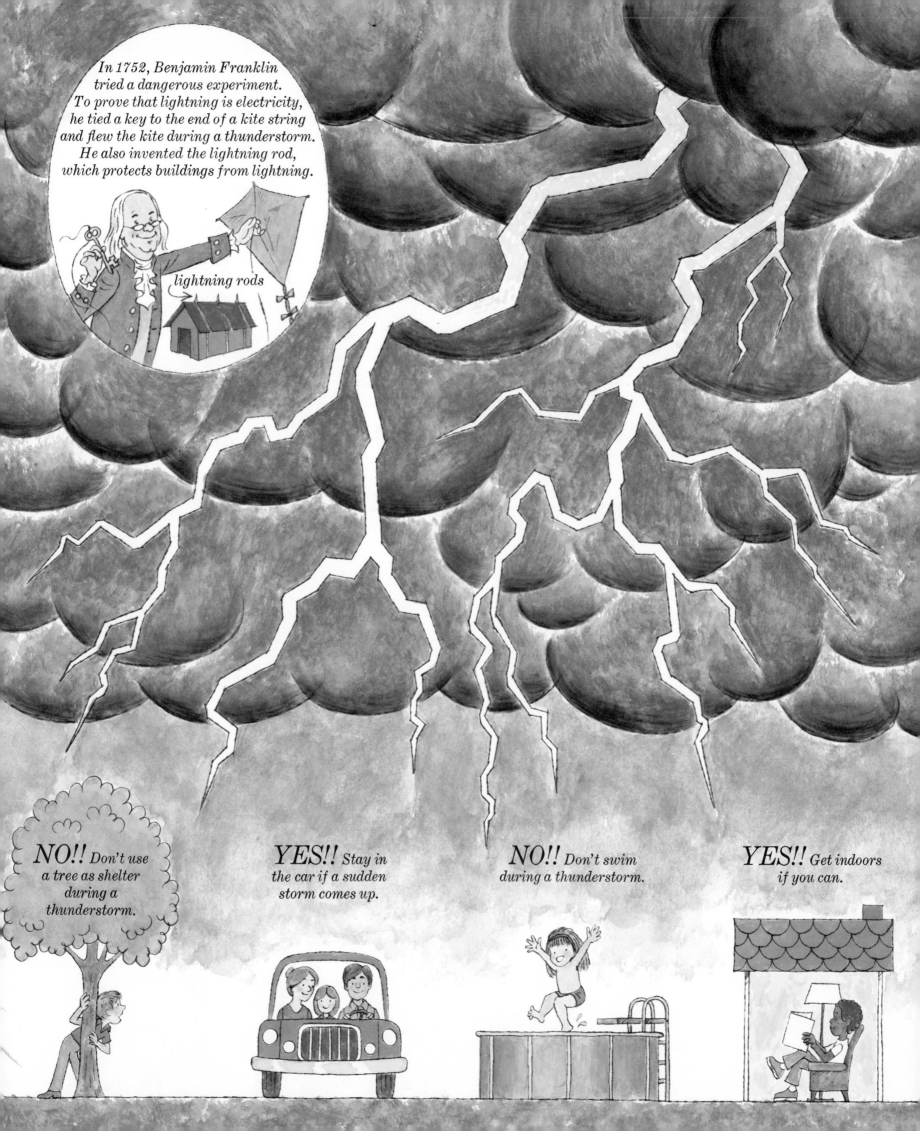

A *STORM is swiftly moving air, usually accompanied by heavy rain or snow.*
The most severe and dangerous storms are tornadoes, hurricanes, cyclones, and blizzards.

A **TORNADO** has the strongest winds of any storm. It is powerful enough to make a house or barn explode or to pick up a car and drop it hundreds of feet away. The spinning wind of a tornado blows 300 miles an hour or even more. As the tornado moves forward, its funnel moves from side to side and sucks up everything under it. Sometimes it will suck up a pond full of frogs and fish and later they'll come raining down out of the sky miles away. Tornadoes have also been known to pluck the feathers from chickens.

Tornado funnels are formed in thunderclouds when hot air rises with unusual violence and cold air rushes in suddenly from all sides and starts to spin. Luckily, tornadoes are usually only about ⅛ of a mile across (compared to a hurricane, which can be hundreds of miles across), so all the terrible damage a tornado does is confined to its narrow path. Also, it usually covers only about 15 miles and is over in about half a minute.

A **TWISTER** is what people in some parts of the United States call a tornado.

It's black up here because of the dust, dirt, and debris that have been sucked up.

The long tornado spout looks like an elephant's trunk.

walls explode outward

roof blows off

heading for the storm cellar

A HURRICANE starts at sea near the equator. All summer long, the sun heats the water and the air above it. Toward the end of the summer or early in the fall, trade winds blowing from opposite directions begin to whirl around each other, causing the warm air rising from the sea to start spinning. It spins high up above the ocean. There the moisture in the air condenses, and heavy rain begins. The air spins faster and the storm grows to 300 or 400 miles across. The storm is called a tropical storm until its winds reach 75 miles an hour — then it's a hurricane.

As the hurricane moves along, huge ocean waves may smash onto beaches, and houses and trees along the shore may be damaged or destroyed. If the hurricane moves inland it gradually loses its power.

A TYPHOON is a hurricane in the China Sea and the western Pacific area.

A CYCLONE is the name for a hurricane that happens in the Indian Ocean.

The center of the hurricane is calm.

Hurricanes rotate at the rate of 75 miles an hour or more.

Hurricanes begin over water in the tropics.

some go north

some go south

A big hurricane may be 500 miles across, a small one only 25.

A BLIZZARD is a heavy snowstorm with winds blowing at more than 32 miles an hour. In a severe blizzard, the temperature is below 10°F (−12°C) and the snow may fall so heavily that you can hardly see your hand before your face.

Blizzards start when freezing polar air meets warm, moist tropical air. Along this meeting line, strong winds begin to blow and snow falls heavily. Often blizzards happen after a period of unusually warm winter weather.

A big blizzard can just about stop most normal activity. Huge drifts may form, some reaching as high as the second story of a building. Schools are forced to close. Telephone lines break under the weight of the snow. Food and other supplies cannot be delivered to the cities by truck or train. Snowplows work overtime. Sometimes many days go by before all roads and train tracks can be cleared.

Warnings of approaching blizzards are broadcast by television and radio, but weather forecasters can't always tell the path a blizzard will follow and how severe the blizzard will be.

Blizzards begin at the North or South Pole.

A blizzard strikes—it's very cold, with strong winds, heavy snow, low visibility.

The blizzard's over—often the piled-up snow stops all traffic. Schools may close.

THE SEASONS are spring, summer, autumn (also called fall), and winter. Each season lasts about three months. Four seasons make a year, the time it takes for the earth to circle the sun. The earth is tilted as it makes its trip around the sun. We have seasons because of this unchanging tilt, for each hemisphere gets different amounts of heat from the sun at different times of the year.

When the North Pole is tipped toward the sun and the South Pole is tipped away, it is summer in the Northern Hemisphere and winter in the Southern Hemisphere. Six months later, when the earth has

AUTUMN is the season when the leaves of many trees turn from green to red, orange, yellow, and brown. They fall to the ground, so some people call this season "fall." It is harvest time. Farmers gather crops they planted in the spring. Birds fly toward the equator to avoid the coming cold. The days grow shorter and the nights longer. Above the equator, autumn begins September 22 or 23 and ends December 22 or 23. Below the equator, autumn begins March 20 or 21 and ends June 21 or 22.

DECEMBER ← NOVEMBER ← OCTOBER

Autumn is over and winter begins December 22 or 23. The Northern Hemisphere is tipped away from the sun.

THE SEASONS IN THE NORTHERN HEMISPHERE

Winter's over and spring begins on March 20 or 21.

JANUARY → FEBRUARY → MARCH →

WINTER is the coldest season of the year. In the parts of the temperate zones near the tropics, it is the rainy season. In the parts of the temperate zones toward the Poles, it's the time for ice skating, skiing, snowballs and heavier clothes to keep you warm. The leaves are gone from most trees, birds are scarce, and some animals sleep all winter. Above the equator, winter begins December 22 or 23 and ends March 20 or 21. Below the equator, winter begins June 21 or 22 and ends September 22 or 23.

moved around to the sun's other side and the North Pole is tipped away from the sun and the South Pole toward it, it's winter in the Northern Hemisphere and summer in the Southern Hemisphere. So the seasons in each hemisphere are the exact opposite of each other.

In the 1500's, people believed the sun circled the earth. Copernicus said Earth circled the sun. An unpopular idea.

SEPTEMBER ← AUGUST ← JULY

Summer's over and autumn begins on September 22 or 23.

Spring is over and summer begins on June 21 or 22. The Northern Hemisphere is tipped toward the sun.

APRIL → MAY → JUNE

SUMMER is the hottest season. In summer, some people are on vacation; they can go to the beach, swim in a lake, or just relax under the shade of a big tree, now full of green leaves. Flowers and plants thrive under the warm sun. Birds are busy raising their new families, and so are most animals. Above the equator, summer begins June 21 or 22 and ends September 22 or 23. Below the equator, summer begins December 22 or 23 and ends March 20 or 21.

SPRING is the time when colder weather is gone and playing outdoors doesn't mean putting on all those heavy clothes. Spring winds are just right for flying kites. Spring is planting time. Trees are getting new leaves, and some flowers are in bloom. Birds that have been away come back. Animals that slept all winter wake up and search for food. Above the equator, spring begins March 20 or 21 and ends June 21 or 22. Below the equator, spring begins September 22 or 23 and ends December 22 or 23.

In all parts of the world, people dress to suit the climate:

N. Polar Zone

Canada Alaska Lapland Siberia Greenland

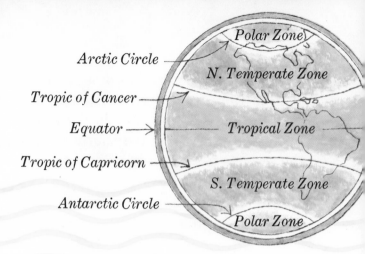

Arctic Circle — Polar Zone
N. Temperate Zone
Tropic of Cancer
Equator — Tropical Zone
Tropic of Capricorn
S. Temperate Zone
Antarctic Circle — Polar Zone

N. Temperate Zone

France Texas Russia Turkey China

Tropical Zone

Tahiti Congo Hawaii Amazon Kenya

S. Temperate Zone

Chile S. Africa Argentina Uruguay Australia

S. Polar Zone

No people down here except a few explorers.

The CLIMATE of a place is the usual weather conditions of that place over a period of years. Earth has five general climate zones: two polar zones (the cold areas), two temperate zones (the mild areas), and one tropical zone (the hot area).

Climate affects the way we dress, build our houses, and plant our crops. People who live in a cold climate wear heavy clothes and build their houses to keep out the cold. They travel by sled, snowshoes, skis, or snowmobile and eat food that gives them extra energy. People who live in a hot climate don't need clothes for warmth, but they may need them for protection from the sun's rays. They build houses that keep out the rain and sun, but let cooling breezes in.

The angle at which the sun's rays strike each zone determines that zone's climate. The sun's rays hit the tropical zone head on, so that the most rays land on each square mile. That's why it's the hottest zone. The temperate zones are mild because the sun's rays strike the surface at an angle, and fewer rays land on each square mile. The polar zones are coldest because the sun's rays strike them at a very slanted angle and the fewest rays reach each square mile.

Each zone has the same four seasons, but summer in the tropical zone is much hotter than summer in the temperate zone. Winter in the polar zone is much colder than winter in the temperate zone.

Climate is more than temperature. It's also the moisture in the air and how much rain or snow falls. Climate depends on how close the area is to water. It also depends on how high above sea level the area is, because the temperature of the air falls as the air gets thinner. General wind patterns also affect climate.

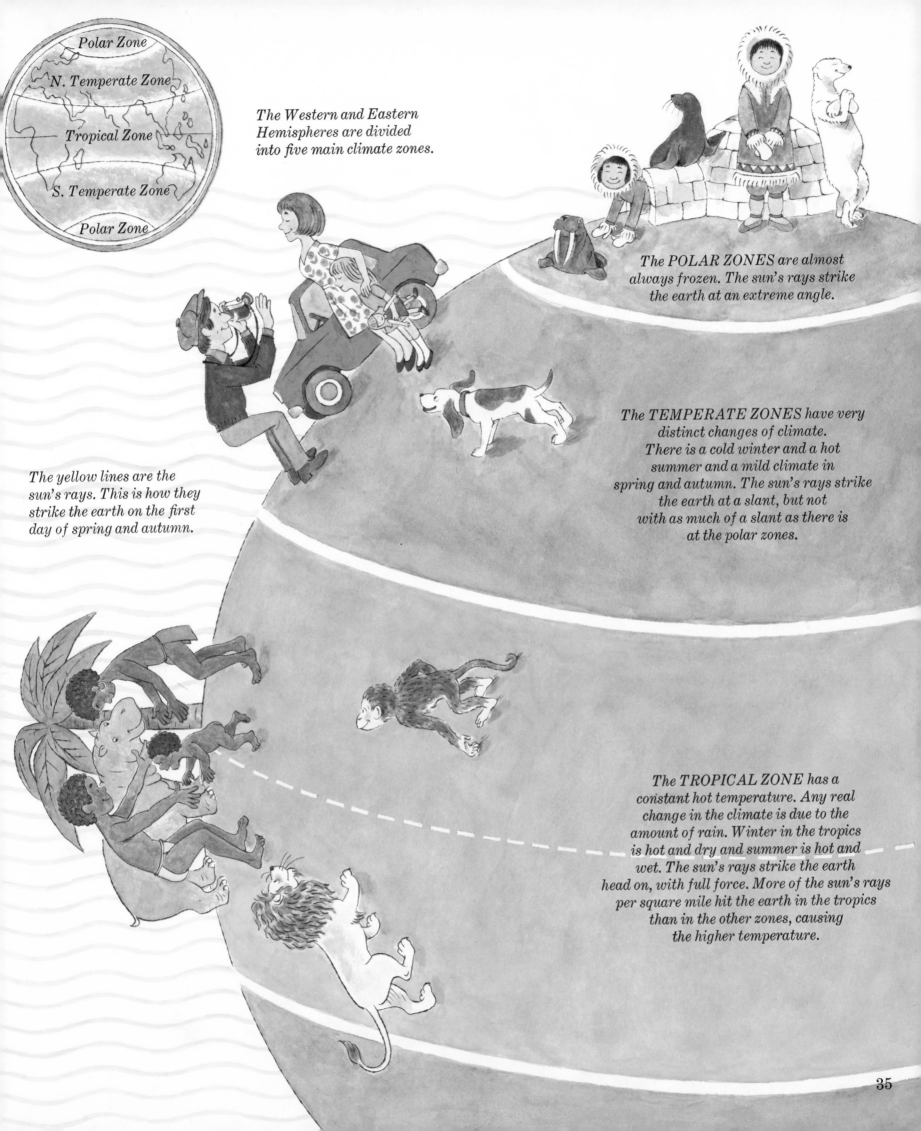

Polar Zone

N. Temperate Zone

Tropical Zone

S. Temperate Zone

Polar Zone

The Western and Eastern Hemispheres are divided into five main climate zones.

The POLAR ZONES are almost always frozen. The sun's rays strike the earth at an extreme angle.

The TEMPERATE ZONES have very distinct changes of climate. There is a cold winter and a hot summer and a mild climate in spring and autumn. The sun's rays strike the earth at a slant, but not with as much of a slant as there is at the polar zones.

The yellow lines are the sun's rays. This is how they strike the earth on the first day of spring and autumn.

The TROPICAL ZONE has a constant hot temperature. Any real change in the climate is due to the amount of rain. Winter in the tropics is hot and dry and summer is hot and wet. The sun's rays strike the earth head on, with full force. More of the sun's rays per square mile hit the earth in the tropics than in the other zones, causing the higher temperature.

35

PLAINS are usually the parts of the earth's surface that are flat and low, or are at least lower than the ground around them.

There are plains of all sizes. Some plains are tiny, measuring only a few acres. Some plains are vast, taking up as much as hundreds of thousands of square miles. The biggest plain of all begins on the border between Spain and France. It stretches across the top of Europe and Asia all the way to the Bering Sea—nearly halfway around the world. The American Great Plains stretch 2,000 miles north and south along the eastern base of the Rocky Mountains.

Some plains began as the beds of oceans or very large lakes. When the bottoms of the oceans or lakes were lifted because of movements of the earth's crust, or when the water level fell, the exposed surface became a plain. Another kind of plain is created by large rivers which deposit rich soil on the areas around them every time there is a flood.

These plains are called flood plains. The city of New Orleans was built on a flood plain made by the flooding of the Mississippi River. Lava plains may form when molten rock from a volcano fills a valley. Glacial plains may be left when a glacier melts.

Almost all areas of the earth have plains. Generally plains are in the center of each continent and along its coasts.

Where there are plains there are usually many people. Generally, plains soil is good for farming, and the flat land makes it easy to use tractors and other farm equipment. The flatness is also perfect for building straight roads and railroad lines. The people of ancient Egypt developed their great civilization on the Nile plain. The famous old cities of Greece were also built on plains, and people have been finding plains good places to live ever since.

Most of our world population, our cities, and our agricultural, industrial, and commercial areas are on plains.

Most of the world's food is raised on plains.

PLATEAUS are large, level areas of land. They are usually over 2,000 feet high, higher than plains. Plateaus are like plains because both of them are flat, but on at least one side of a plateau there is a steep slope downward, and often the other sides rise up into hills or mountains.

Streams and rivers often run through plateaus, wearing away the soil and creating valleys that may deepen into spectacular canyons. The Grand Canyon was formed this way. Plateaus are useful to sheep and cattle raisers. American cowboys often use them as grazing lands for their cattle.

Sometimes people refer to a plateau as a "tableland," because the plateau's flat top stands up like a table from the land around it.

Most plateaus were formed by the movements of the earth's crust, which pushed up large flat areas of land. Others were formed by huge amounts of lava pouring out of a volcano and spreading out over the surrounding land, creating a high flat area.

HILLS are usually areas of land rising a few hundred feet above their surroundings. Most hills have a rounded appearance and are small compared to mountains. But there are exceptions: in South Dakota, there is a hill that rises to more than 7,000 feet, higher than many mountains. There is no limit to how high a hill is and no agreement on the difference between a hill and a mountain. What some people call high hills, others call mountains.

Some hills were formed when high mountains were slowly worn away to their present size and shape. Other hills were once plains with streams or rivers running through them. The streams or rivers eroded the plains and created valleys. The parts that didn't wear away were left behind as hills. Sand and stones left by glaciers also formed some hills.

Hills usually have rounded tops.

Some plateaus are good grazing grounds.

Mountains are created in different ways by movements of the earth's crust, caused by forces inside it.

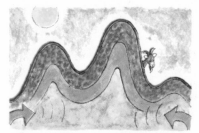

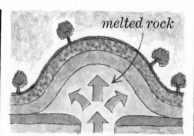

melted rock

FOLDED mountains rise up as pressure from inside the earth squeezes land into folds.

BLOCK mountains rise at a fault (break). One side of the fault is pushed up high.

fault

DOME mountains rise up when melted rock below the surface causes it to bulge.

VOLCANIC mountains form when boiling lava spurts out, piles up, cools, then hardens.

Mountains are worn down gradually by many forces from the outside.

glacier

Rushing water, with sand and stones, wears down the most rugged mountains.

Wind-blown sand chips away at the mountains and slowly erodes them.

Glaciers move down mountains, pushing rocks and soil, and carving deep valleys.

MOUNTAINS cover about a fifth of all the earth's land area. Although mountains look permanent and unchanging, they are not. They rise up, then wear away, and eventually disappear. When a mountain is young (a mere million years old), it usually has sharp, jagged peaks and steep slopes. Then water, grinding glaciers, wind, and blowing sand begin to wear it down. When the mountain has grown old (hundreds of millions of years later), its peaks are worn down and rounded off.

Mountains are created by movements inside the earth's solid rock crust. Under the crust, inside Earth's mantle, hot and sometimes molten rock and metals are in constant motion. They may squeeze the earth's crust and push it up, making folded mountains. They may crack the crust and push part of it up to make fault-block mountains. Or molten rock may push up from below and make the crust

bulge into dome mountains. All of this happens very slowly, over a time span of millions of years.

Volcanic mountains are the only mountains that grow fast, in a matter of a few hundred years. They are formed when lava, molten rock from the mantle, spurts out through the crust and builds into a mountain. Sometimes a volcano will start and shoot up very quickly. Not too long ago in Mexico, a new volcano started. Boiling rock began to pour out of the ground and pile up. In only nine years, it was 1,600 feet high!

Mountains exist at sea, too. There is a long series of mountain ranges at the bottom of the oceans, longer than any on the continents. These mountains sometimes stick out of the water. If your home is in Iceland, or on one of the Hawaiian Islands, or the West Indies, you are living on the top of an underwater mountain.

Altitude has an important influence on climate. The weather is usually warmest at the base of a mountain, cooler on the slopes, and coldest at the peak.

MOUNTAIN CLIMBING has been a sport for only about 200 years. Before that people just didn't climb mountains for fun. They thought climbing was difficult and dangerous, and they saw no reason for it.

Mountain climbing began in France in 1760 near Mt. Blanc, in the Alps. A naturalist named Benedict de Saussure went there to collect plants that grow in the Alps. He imagined the thrill of looking down from Mt. Blanc and offered a prize to the first person to climb it. Twenty-six years went by before two men finally won the prize.

One by one the great peaks of the world have been scaled—even Mt. Everest, the highest, in 1953. Today climbing isn't only a sport, it's also a means of gathering scientific information.

Mt. Blanc is 15,771 feet high.

Mt. McKinley is 20,320 feet high.

Dr. Michel Paccard and Jacques Balmat were first to climb Mt. Blanc, the highest Alps peak. Paccard, a French doctor, had tried many times to climb it, but failed. Then he planned a new route. He couldn't afford a guide, but Balmat, his porter, offered to help. They started out one morning in 1786. The climbing was awful, with more ice and deeper crevasses than before, but they succeeded and won the de Saussure prize.

Hudson Stuck, the Archdeacon of the Yukon, and his three companions were the first to climb Mt. McKinley in Alaska. Many had tried—a group of goldminers had reached McKinley's second highest peak, but no one had ever reached the top. Stuck prepared for the climb carefully, picking the right men and sending for the best equipment. In June of 1913, after a great struggle, the group looked down from the tallest mountain in North America.

Mt. Blanc is on the border between France and Italy.

Mt. McKinley is in Alaska.

This is some of the equipment mountain climbers use:

the all-important climbing rope

Pitons go into cracks to hold rope.

oxygen tank and mask

Crampons fasten to shoes for ice-walking.

an ice ax, a big help to a climber

Mt. Everest is the tallest mountain on Earth, 29,028 feet high.

Mt. Huascaran is 22,204 feet high.

Annie Smith Peck, an American professor of Latin, climbed Mt. Huascaran's highest peak in 1908. This was the first major peak to be climbed first by a woman. She had failed on two earlier attempts. This time, overcoming steep, icy slopes, bitter wind, and freezing cold, she and two Swiss guides made it. Miss Peck vanquished many other famous peaks, and she was still climbing mountains when she was 82 years old.

Mt. Huascaran is in Peru. —⟶

Edmund Hillary and Tenzing Norgay were first to climb Mt. Everest, the world's highest mountain. Hillary, a New Zealander, had taken part in other British attempts at Everest, as had Norgay, a Sherpa from Nepal. This time they set out with a well-equipped British expedition. Hillary and Norgay made the final climb to the top on May 29, 1953. Queen Elizabeth knighted Hillary and awarded Norgay the George medal.

Mt. Everest is on the border —⟶
between Nepal and Tibet.

41

The green leaves absorb air, mix it with water, and turn it all into sugar (tree food). The sun provides the energy.

a tree's complete diet: air, water, and minerals (plus sunlight)

The roots absorb water from soil and send it up through the trunk.

Trees that lose their leaves in autumn are called deciduous. Trees that keep their leaves all winter are called evergreen.

maple

ash

Trees that lose their leaves in autumn have broad leaves. Some are one-piece simple leaves. Some are compound leaves, each with many leaflets.

birch

Most evergreen trees have thin, needle-like leaves.

pine

Each year a tree adds a ring of new wood to its trunk. To tell a cut-down tree's age, count the rings.

A **TREE** is a plant, and like most plants it has roots, a stem (its trunk), and leaves, flowers, and fruit. Like most plants, trees make their own food. The roots absorb water and minerals from the soil and send them up to the leaves. There the water and minerals are combined with carbon dioxide from the air and are formed into sugar, which is food for plants. The sun's energy makes this process happen and is also stored in the sugar. So trees and other plants can live without moving from place to place. Other living things are always running, crawling, hopping, flying, swimming, or shopping in search of food, but plants make their own.

Trees are usually much bigger than other plants. California Redwood trees, the tallest things alive, are more than 350 feet tall. Trees also live much longer than any other living thing, plant or animal. Some trees are more than 4,000 years old.

A tree has three parts—the roots, the trunk, and the crown (all the branches and leaves). The roots spread out underground, anchoring the tree firmly in the soil. They take up more space below ground than the branches do up above.

The trunk supports the crown of the tree and transports the water and minerals from the roots up to the leaves through a layer of tiny tubes. Food made by the leaves goes back down the tree in other tiny tubes in a layer of inner bark. If the inner bark is cut through all the way around the trunk, the tree's food flow is cut off and it will die of starvation. Each year, a new layer of wood grows inside the inner bark. When a tree is cut down, we can count these rings of wood growth and tell its age.

In the crown, branches stretch outward. From the smaller branches grow the twigs that hold the leaves (where the tree's food is made), flowers, and seeds. We don't notice the flowers on some trees, but every tree has them. Flowers make seeds, and when the seeds fall to the ground, new trees may grow from them.

Wood from trees is used to make many things . . . furniture, bats, toys, pencils

Here is just a small selection of the world's great variety of trees:

fruit

leaf

APPLE *trees give us one of our most popular fruits. There are hundreds of kinds.*

SPRUCE *trees, tall, pointy evergreens, love cold weather. Many wind up as paper.*

MAPLE *trees provide great syrup. The wood is hard, the fall leaves brightly colored.*

CHESTNUT *trees have lots of flowers in the spring. The nuts are in spiny husks.*

PALM *trees haven't any branches. Huge leaves come out of the top of the trunk.*

OAK *trees have tough, hard wood. They grow from acorns, nuts wearing little hats.*

PINE *trees have needles fastened together at one end, two to five in a bunch.*

ELM *trees' trunks divide into very graceful branches about a third of the way up.*

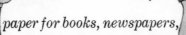

houses,   rackets, fabrics for clothes,   paper for books, newspapers,   and gift wrapping . . . and trees give us fruit!

FORESTS are large areas of land covered with trees growing close together. (Woods are smaller areas of tree-covered land.) One third of the world's land surface, or 7½ billion acres, is covered with forests.

Forests supply wood for building houses, for making furniture, and for all other wood products. Wood is also made into paper, cellophane, and synthetic fabrics which are made into clothes.

Forests are also important because they store water. Forest soil, kept by tree roots from blowing or washing away, holds about 50 times as much water from rain and snow as soil without trees. Eventually the water seeps into streams which are often the source of community water systems.

Forests grow where there is at least 15 inches of rain a year and where there isn't continual frost. There are no forests on high mountaintops above what is called the timberline because it's too cold up there for seeds to grow.

Forest fires can be one of the biggest enemies of the forest, along with insects and tree diseases. Some

Adder	Bear	Cardinal	Daisy	Elk	Fox	Grackle	Hare	Insects	Jay	King snake	Lily	Mole

Seen in the forest:

fires start when careless people leave their camp-
fires or cigarettes burning.

Many animals, birds, and insects live in forests.
In the past, many people lived in the forest and
hunted forest animals for food and leather. Now
people enjoy forests for hiking, fishing, and camp-
ing. Most national parks have hiking trails and
campsites where families can set up tents and have
fun in the outdoors.

Foresters are scientists whose work is to protect
and preserve the forests.

BE CAREFUL WITH FIRE.
BUILD ALL FIRES IN SAFE PLACES.
NEVER LEAVE A FIRE UNATTENDED.
BEFORE YOU LEAVE PUT FIRE OUT BY
SPRINKLING IT WITH LOTS OF WATER.
THANK YOU

| Nuthatch | Owl | Porcupine | Quail | Raccoon | Squirrel | Turtle | Ursinia | Violet | Weasel | Xerus | Yellowbird | Zinnia |

Vines hang from the trees. They're great to swing on!

Armadillos, lizards, turtles, snakes, frogs, and tapirs are all at home in the jungle.

46

There are chimpanzees and elephants in the African jungle, tigers in the Asian jungle, jaguars in the South American jungle.

Monkeys, parrots, big-beaked toucans, and owls live high in jungle trees.

Slow-moving sloths hang upside down and even sleep that way.

JUNGLES are tropical rain forests where more than 100 inches of rain falls each year. Jungle areas stretch around the earth at or near the equator. Almost half the earth's forest lands are jungles.

With all that rain and heat all year long at the equator, vegetation grows fast. The jungle is crowded with trees stretching upward toward the sun. Vines tangle through the trees and hang down. The floor of the jungle is shady and dark—few plants grow there because very little sun reaches down. High above the ground on the trunks and branches of trees, unusual air plants grow without soil. Some air plants have developed ways of catching and storing rainwater. Orchids are air plants.

There are more ants in jungles than any other kind of living thing. One kind of ant, the Army Ant, marches out in company like a battalion of soldiers to search for food. Army Ants overwhelm and eat any living thing in their path, even small animals.

Some of the other insects that live in jungles are butterflies, mosquitoes, wasps, beetles, and termites.

Up in the trees live brightly colored, noisy birds: parrots, macaws, toucans, red chatterers, crested flycatchers, and red-cheeked woodpeckers.

Jungles are the home of snakes of all sizes, including the anaconda, the world's biggest snake, which may grow to be 40 feet long and weigh 600 pounds. The anaconda squeezes its prey to death and swallows it whole, without chewing. Lizards, crocodiles, and turtles are other reptiles found in jungles.

There are also tigers, jaguars, and panthers, monkeys, chimpanzees, and gorillas, tapirs, armadillos, and—with all those ants—anteaters.

All the world's jungle areas are located near the equator.

An anteater follows some ants. They carry bits of leaves like umbrellas.

The air in the jungle is steamy and there's lots of rain—almost 100 inches a year.

47

Swift horses are used in the desert, but they need water often and their hoofs sink in the sand.

The camel is the most useful animal for traveling through the desert.

An oasis is a spot in the desert where there is water. The ground is moist, plants and trees grow.

Desert animals can survive on very little water.

a rattlesnake

a horned lizard   a kangaroo rat   a roadrunner   a desert tortoise

These cactus plants and other desert plants also need very little water.

There are oil fields in the desert now, and many modern vehicles, too.

It can go for days without a drink, can carry heavy loads of up to 1,000 pounds, and its big, wide feet are just right for the soft sand.

DESERTS are large land areas with very little water. Deserts cover a fifth of the earth's surface. Some deserts are always hot; others are hot in summer and cold in winter. Some deserts have dunes; others, rocky hills and mountains. Some have oases (fertile areas with some water); some do not.

Deserts have little water for many reasons, but mainly because they get little rain, only a few inches a year. Some deserts are too far from any large body of water for its rain clouds to reach them. Others are near an ocean, but mountains in between cause clouds to drop their rain before they reach the desert. There are deserts on coasts in areas where the wind always blows away from land, so no moisture can come inland.

Most of the continents have deserts: Africa has the Sahara; Asia, the Gobi; North America, the Mojave; Australia, the Great Sandy; South America, the Atacama.

There are many desert animals, birds, and insects, and all are adapted to life in desert conditions. The camel is an example, storing fat in its hump and absorbing the fat for nourishment when there is no food to eat. The camel's eyes are protected from blowing sand by long lashes and overhanging lids, and its nostrils can snap shut to keep out the blowing sand. The camel also drinks enormous amounts of water and doesn't lose it by sweating.

A variety of plants grow in the desert and get along with almost no rain. Their roots go deep down for water. Their thick leaves and stems store water, and some are wax-coated, preventing evaporation.

Some desert soils are surprisingly fertile and, with irrigation, make good farmland.

People who live or work in the desert need special clothes and other protection.

In the Sahara, camel drivers and other people wear clothes that shield them almost completely from the blazing sun.

A cowboy wears a hat with a wide brim to ward off the hot rays of the desert sun.

This isn't a bad cowboy. His face is covered to keep out the desert sand and dust.

An African bushman in the Kalahari Desert oils his body then lets dust cover it to protect him from the sun.

A Viking, Eric the Red, discovered and settled Greenland A.D. 986.

an igloo—a temporary house made of snow

a kayak—covered with sealskins

Eskimos have been influenced by modern ways, but many still like to live as they always have.

In 1909, Robert E. Peary, with Matthew Henson and 4 Eskimos, was first to reach the North Pole.

seal hunting through the ice

The **POLAR ZONES** are the frozen parts of the earth. The Arctic, which surrounds the North Pole, is an icy ocean encircled by land; the Antarctic, which surrounds the South Pole, is ice-covered land encircled by icy ocean.

Almost all the Arctic Ocean is covered with a thick layer of ice. During the summer, some ice melts at the edges and icebreakers can go through for a distance, but no ship can cross the Arctic Ocean.

The parts of Arctic land not covered with ice are called tundra, and are flat and marshy. The tundra has no trees, but there are grasses, plants, birds,

animals, and insects. About 400,000 people are native to the Arctic, including about 60,000 Eskimos.

During the 6 months that the Northern Hemisphere is tilted toward the sun, the North Pole always has sunlight, day and night. During the 6 months the Northern Hemisphere is tilted away from the sun, the North Pole is always dark. The South Pole also has 6 months of daylight and 6 months of darkness.

The North Pole is as far north as you can be, because no matter which way you go away from it, you will be going south. If you leave the South Pole,

Captain James Cook, first to explore the Antarctic—1778

85% of an iceberg is below water.

seals

Roald Amundsen, using dogs, was the first to reach the South Pole—December, 1911. Robert Scott, who had used ponies instead, got to the pole only one month later.

The female Emperor penguin lays one egg. The male holds the egg on his feet and warms it under his belly.

egg →

Most of the Arctic Region is frozen sea with land around it.

Soviet Union

Europe

North Pole

Alaska

Greenland

tern

owl

duck

goose

caribou

walrus

seal

whale

musk-ox

polar bear

some of the many birds and animals of the Arctic

wolf

The Arctic is rich in minerals and oil—a pipeline brings oil from Alaska to the rest of the U.S.A. The Arctic has weather and other research centers.

no matter which way you go, you will be going north.

Antarctica, the ice-covered continent at the South Pole, is huge—about 1½ times the size of the United States. Antarctica is a high plateau averaging 6,000 feet above sea level (the highest for any continent), with several mountain ranges rising up to 16,000 feet high, and an active volcano 13,200 feet high.

No one has ever seen the land of Antarctica. It is covered with ice—thousands of feet of ice, more ice than exists anywhere else in the world. At the South Pole the ice is 1½ miles thick. Only a few mountain peaks stick up out of it. If Antarctica's ice should suddenly melt, the oceans all over the world would rise 200 feet and drown the coastline cities.

Antarctica is the coldest place on Earth, with temperatures as low as 129°F below zero (−89°C). About 10 inches of snow falls each year (never any rain) and stays there.

The ocean surrounding Antarctica is the world's roughest, with storm winds of up to 200 miles an hour. The waters are full of whales, seals, crabs, and fish. Penguins live along the coast and depend on the ocean for food, because almost nothing edible grows on land. Antarctica has no native people.

Sperm Whale

Blue Whale

Killer Whale

There are many whales in the Antarctic.

South America

South Pole

Antarctica

Australia

Africa

Antarctica is an icy continent with an icy ocean all around it.

Today scientists from many countries are in the Antarctic studying the weather, the stars, the ocean, earthquakes, gravity, magnetism, and what is under all that ice.

If the continents were pieces of a jigsaw puzzle . . .

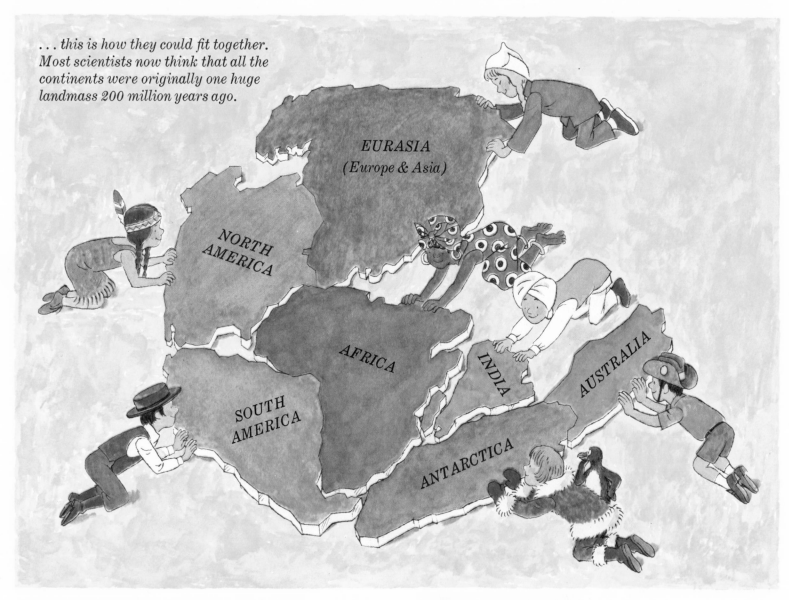

. . . this is how they could fit together. Most scientists now think that all the continents were originally one huge landmass 200 million years ago.

EURASIA
(Europe & Asia)

NORTH AMERICA

AFRICA

SOUTH AMERICA

INDIA

AUSTRALIA

ANTARCTICA

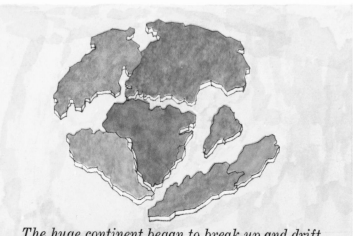

The huge continent began to break up and drift apart. This is the earth 135 million years ago.

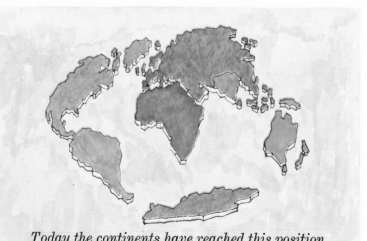

Today the continents have reached this position. They are still moving—1 to 6 inches a year.

The idea of **MOVING CONTINENTS** is one that most scientists now accept. They believe that 200 million years ago the earth was covered by a giant ocean of water with one giant continent of land sitting in it. Then the one continent broke in two, and broke again, until the earth's crust was split into about 20 pieces, called plates. The plates are held up by a layer of hot, denser matter. Some of these plates from the old giant continent became

Madagascar

Africa

the seven continents we know today. Others were covered by ocean. The plates drifted until the continents reached their present locations, and they are still moving, 1 to 6 inches a year.

When moving plates meet, their edges may crumple up into mountains. The motion of the plates also causes earthquakes and volcanic action at the plates' edges.

The idea of moving continents came about when people noticed that South America and Africa would fit together neatly like jigsaw puzzle pieces. In 1912, the scientist Alfred Wegener said all the continents would fit together and that ages ago they were all joined as one continent. He had some traces of remains of ancient animals, fish, insects, and plants to prove his theory, but didn't know just how the continents were able to move. Now the idea of drift-

ocean plate pushing under the land plate, creating mountains

Andes

South American plate

Pacific plate

ing plates explains this movement and scientists have gathered much more evidence that the continents once were joined.

Here are some recent findings. The hippopotamus lives in Madagascar and also on the coast of Africa.

250 miles of water lie between the two—too far for a hippopotamus to swim. So scientists think Madagascar and Africa were once joined. Further evidence is that the remains of a dog-sized reptile called Lystrosaurus, previously found in South Africa and India, have recently also been found in Antarctica. This indicates that the three continents must have long ago been parts of one continent.

An **EARTHQUAKE** is a sudden shaking or trembling of the ground, set off by a break or slipping of the rocky crust below the earth's surface. An earthquake usually lasts for only a few seconds, but shock waves travel out in all directions. Buildings near the break or in the area of the shock waves may be damaged or destroyed.

About a million earthquakes happen every year.

Most of them are weak or occur under water or in remote places. Only about 100 cause trouble.

There are certain areas around the earth where nearly all earthquakes take place. Those are the same areas where the large plates of the earth's crust meet. Scientists believe that the stress that exists where plates meet causes earthquakes.

A **VOLCANO** is an opening in the earth's crust through which molten rock (lava), gases, and pieces of rock come bursting out. The lava and rock pile up around the opening into the shape of a cone. If a volcano keeps erupting from time to time, this cone may grow into a high mountain. An active volcano can be very dangerous, threatening lives and property. Volcanoes usually happen in the same unstable areas of the earth's crust as earthquakes.

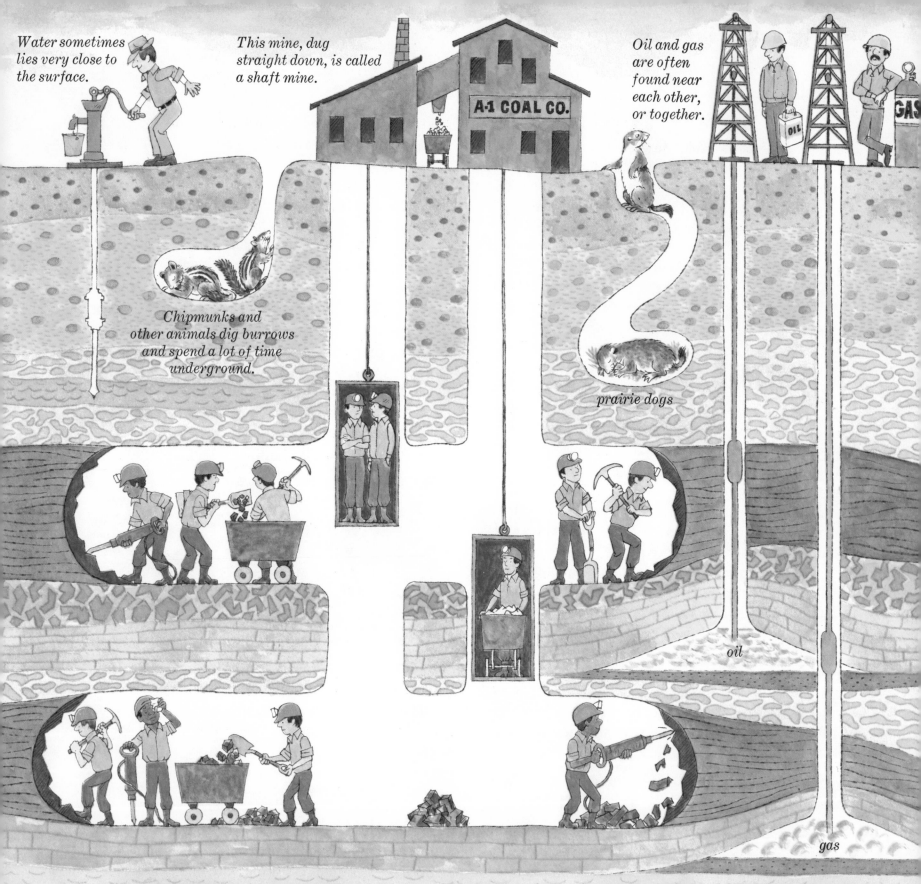

Water sometimes lies very close to the surface.

This mine, dug straight down, is called a shaft mine.

A-1 COAL CO.

Oil and gas are often found near each other, or together.

GAS

OIL

Chipmunks and other animals dig burrows and spend a lot of time underground.

prairie dogs

oil

gas

Buried UNDERGROUND are some of the materials people value most: coal, oil, and gas for heat, cooking, transportation, and making electricity; and metals for making tools, cars, planes, ships, bridges, buildings, and coins. There are also diamonds and other jewels, and salt and water.

People first used metals found on the earth's surface for making utensils, weapons, and ornaments. Even meteorites that fell from space were used for early iron tools. Eventually people learned that

there were veins of metal ore underground. The first miners dug holes and hoisted the ore up in baskets. The deepest modern mine, in South Africa, is 12,600 feet deep. The temperature at that depth is 131°F (55°C), so the mines must be air-conditioned.

Petroleum, first noticed because it seeped out of the ground, was used for centuries as medicine by the Egyptians, the Chinese, and the American Indians. In 1850, it was discovered that petroleum could be used instead of whale oil to light lamps.

54

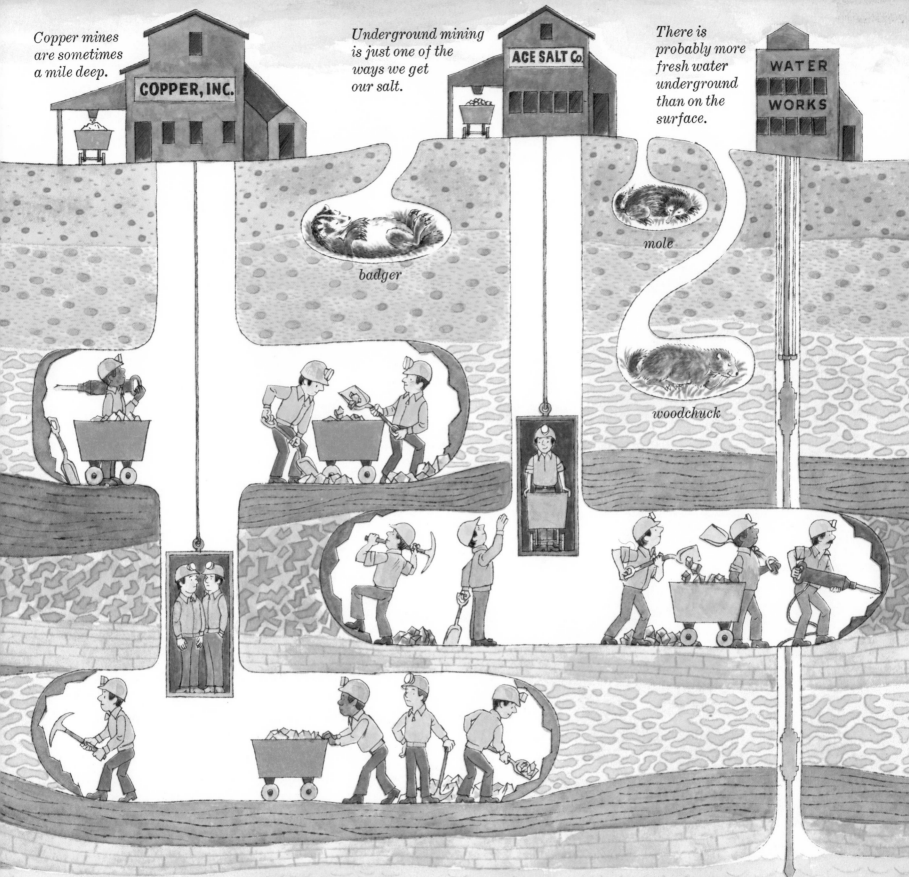

Copper mines are sometimes a mile deep.

COPPER, INC.

Underground mining is just one of the ways we get our salt.

ACE SALT CO.

There is probably more fresh water underground than on the surface.

WATER WORKS

badger

mole

woodchuck

People rushed to find it and soon pumps were pumping it out of the ground.

There are huge mines in almost every country on Earth today. There are oil and gas wells, too, spread all over the world. People even drill for gas and oil in the rocky crust beneath the ocean.

There's water almost everywhere under the earth's surface. Water from rain or snow that doesn't run off into streams sinks down through the soil until it reaches a watertight layer of rock. It collects above the rock. A pump can bring that water up if the water is close enough to the surface. Whole cities, far from lakes or rivers, rely on pumped-up underground water.

Under the Sahara Desert and under parts of the American Southwest lie huge amounts of "fossil water," which is water left over from the Ice Ages.

Miners are not the only underground workers. Earthworms, ants and other insects, and some animals live and sometimes find food there.

55

Early people found that caves offered safety from wild animals and shelter from rain and cold. They lived near the entrance, where

Centuries ago, some American Indians in the southwest built easy-to-defend towns in big caves high up in cliffs.

Mama, am I a troglodite?*

Oui, ma petite!

* A troglodite is anyone who lives in a cave.

In France today, there are people who still live in little villages which have been built completely inside caves.

CAVES are naturally hollowed-out spaces, sometimes in the sides of rocky hills and mountains, sometimes underground. Many caves are made of limestone, a soft stone that dissolves easily. A limestone cave is formed when surface water seeps down through cracks in the limestone, dissolves it, and carries it off, hollowing out a cave underground. As more water seeps down some of it may dry on the ceiling, leaving bits of the mineral calcite, which makes up limestone. More water collects on these bits, leaving more calcite. In this way, the bits grow into stalactites hanging down from the ceiling. Other drops of water with calcite fall to the floor and dry, leaving bits of calcite on the floor. These become stalagmites growing up from the ground. Some stalactites and stalagmites meet in the middle and form stone columns or curtains.

Sometimes caves are made by waves which beat constantly against high, rocky coastlines and wear holes in them. Sometimes the surface lava flowing out of a volcano cools first, and the lava underneath flows out from under it. This makes hollow spaces, or lava caves. In glacier areas, ice sometimes freezes with hollow spaces inside it, forming ice caves.

Bats, who like the dark, often live in caves.
They are the only mammals that can fly.

smoke from their fires could escape and where it was light during the day. On the walls they painted pictures of animals they hunted.

Animals, birds, fishes, and insects live in caves. Some bears and snakes use caves as places for their winter hibernation. Other animals live in caves at night but go outside each day. Others never leave the cave, and these have changed over many, many generations, and now are totally blind. Bats are not blind, as some people suppose, but need their built-in sonar to fly safely in the dark. A flying bat makes high-pitched twittering sounds and the sounds hit objects in its path, bounce back, and warn the bat to avoid the objects.

Caves were probably the first indoor homes for people. In caves where early people lived, scientists have found stone tools and lamps. Bones of animals that no longer exist, such as mammoths and wooly rhinoceroses, have been found in caves where early hunters left them.

The scientists who explore and study caves are called speleologists. The people who explore caves as a hobby are called spelunkers. Careful spelunkers wear safety helmets, warm clothing, and high waterproof boots, and they carry lamps and ropes. They *never* go into a cave alone and without telling someone else where they are going.

Limestone caves form when water seeps down from the surface through cracks in the stone. Water is slightly acid from contact with air and the soft limestone.

The rainwater begins to dissolve the limestone along vertical and horizontal cracks. The cracks become bigger, especially where they join and cross each other.

The openings become still wider, and finally a cave is formed. This takes hundreds of thousands, maybe even a few million, years.

Remember: If it clings to the ceiling—
it's a STALACTITE.
If it grows from the ground—
it's a STALAGMITE.

57

ROCKS make up almost the entire earth. Although most of the earth's surface is covered with water, the layer of water is only 6 miles deep at its deepest. That isn't much compared to the earth's total diameter of about 8,000 miles.

Earth's rock crust lies underneath the ocean beds and the land's surface. We aren't aware of it because it is covered by the ocean's water or the land's soil. Actually, the soil itself is rock, broken down to very fine particles.

Beneath the earth's outer rock crust is the mantle, which is a layer of hot and sometimes molten rock. The mantle makes up 80% of the earth. Inside the mantle is the earth's very hot core.

Geologists, who study the rock of the earth's crust, learn about the history of the earth, a history that goes back billions of years. In the rock they can see the remains of entire mountain ranges that have risen up, crumbled, and worn away. Near oceans, stones on the beach tell of big cliffs that once were there. Traces of ancient sea beds lie in rock that now is part of high mountains. Great stones, marked with deep grooves from ice and stones moving over them, are evidence of the advance and retreat of glaciers during the ice ages.

Also hidden in rock are fossils that tell of plants and animals that once were alive on the earth and now are gone forever. Nothing would be known of the dinosaurs if it weren't for this fossil record in the rocks.

HOW TO BE A ROCK COLLECTOR also known as a "ROCK HOUND"

Collecting rocks can be a lot of fun. This equipment would be very useful.

pad and pencil
bag
pocket-knife
chisel
hammer
magnifying glass

A beach is a good place to look for . . .
rocks! stones! pebbles!

These minerals give us some important metals.

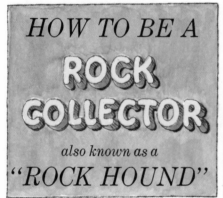

Cinnabar gives us mercury. *It is used in thermometers.*

Argentite yields silver, *used for all kinds of silverware.*

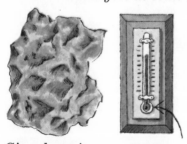

Cassertite is the source of tin, *used to coat tin cans.*

Bauxite yields aluminum, *used for cooking utensils.*

MINERALS are what rocks are made of. They occur naturally in the earth and are made of crystals.

Most rocks are more than one mineral, but some are just one. A piece of granite is made of quartz, feldspar, and mica, but a piece of pure sandstone is all quartz and no other mineral.

Some minerals are harder than others. The hardest one is diamond and the softest is talc. There is a simple scratch test for mineral hardness: a mineral can scratch another that's equally hard, or softer, but not one that's harder. A diamond can scratch all other minerals but no other mineral can scratch a diamond except another diamond. Scientists use the scratch test in classifying minerals. Some other things they look for are the weight and color of minerals, the way they break, their shininess, and what shape crystal they have.

There are three basic groups of rocks. The groupings are determined by how rocks are formed.

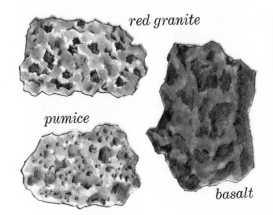

red granite

pumice

basalt

IGNEOUS ROCKS
form when hot, melted rock from inside the earth cools and hardens.

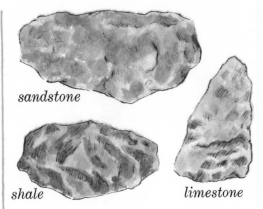

sandstone

shale

limestone

SEDIMENTARY ROCKS
are bits of solid or dissolved rock squashed by pressure into new rock.

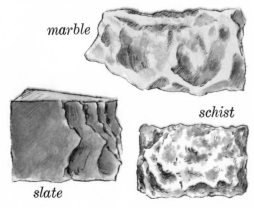

marble

schist

slate

METAMORPHIC ROCKS
form when buried rocks are changed by heat and pressure into new rock.

Look for rocks in your backyard, or try the base of a mountain.

If you live near a quarry, ask permission to visit and pick up some rocks.

and now . . . some very nice collections

CRYSTALS are what minerals are made of. All crystals have geometric shapes. The geometric shape of a crystal is the exact copy of the unseen arrangement of very tiny particles, called atoms, which make up the crystal. There are many kinds of crystals, but only six basic crystal shapes.

Salt, which is a mineral, is made up of crystals. If you look at salt through a magnifying glass, you'll see that all the little crystals are shaped like small cubes. The arrangement of the atoms that make up each salt crystal is also cubic. Sugar is also made up of crystals, and their shape is revealed under a magnifying glass. Calcite, the mineral from which stalactites and stalagmites are formed, has flat six-sided crystals.

All minerals are made up of crystals. Some crystals are more perfect in their natural shape than any stone cut by a master gem cutter.

Crystals have 6 basic shapes. Here's an example of each:

Pyrite is <u>cubic</u>.

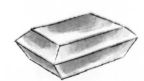

Sulfur is <u>orthorhombic</u>.

Zircon is <u>tetragonal</u>.

Gypsum is <u>monoclinic</u>.

Rhodonite is <u>triclinic</u>.

Quartz is <u>hexagonal</u>.

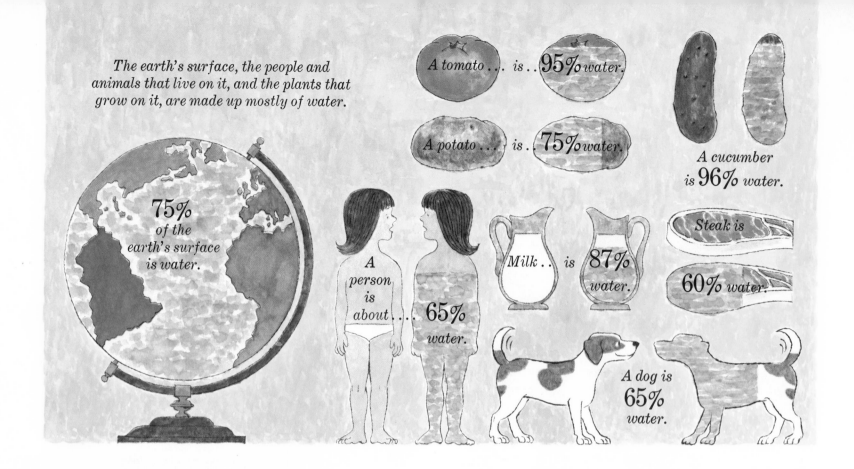

The earth's surface, the people and animals that live on it, and the plants that grow on it, are made up mostly of water.

A tomato... is.. 95% water.

A potato... is.. 75% water.

A cucumber is 96% water.

75% of the earth's surface is water.

A person is about.... 65% water.

Milk.. is 87% water.

Steak is 60% water.

A dog is 65% water.

WATER covers most of the earth's surface. Of the earth's water, 97% is the salty water of the oceans. The other 3% of the earth's water is fresh (not salty). However, most of the fresh water is frozen and remains in the areas around the North and South Poles and on mountaintops, and is not presently available for use. Only 1% of all the earth's water is fresh and not frozen. Half of that 1% seeps underground and supplies water for plants and trees, and the rest is left for us to share with the other land animals. All of the great rivers and lakes, the streams and ponds and underground springs that provide water for drinking, cooking, washing, and all our other needs make up less than ½% of Earth's water.

The fresh water we drink and the ocean's salt water are actually the same water. When the sun heats the ocean's surface, causing water vapor to rise and form clouds, the salt is left behind in the ocean. (In this way, too, other impurities are cleaned from the water.) The ocean doesn't become more and more salty as evaporation goes on because fresh water is always flowing back into it and salt-free rain is always falling back into it. Even when

We use water at home, in industry, for irrigation, for fun. Here are some familiar water uses:

drinking

cooking

canoeing

putting out fires

car washing

showering

toilet-flushing

swimming

We have as much water on Earth as we ever had. We use the same water over and over again and never lose any part of it. The same water was here

...a million years ago,

a thousand years ago,

and is still here today.

1. *Heat from the sun evaporates the ocean's surface.*

2. *The invisible vapor rises from the ocean, then forms into clouds.*

3. *When rain falls from clouds only part of it falls over land; most falls back into the ocean.*

4. *Rain that falls on the land returns to the ocean through streams, rivers, and underground ways, to start the cycle again.*

THE WATER CYCLE: how the oceans' waters circulate over and over again.

rain or snow falls on land, all the water, every bit of it, will eventually return to the ocean. This cycle continues endlessly—the sun's heat makes vapor rise from the ocean, the vapor condenses into clouds, the clouds shed rain or snow that falls into the ocean or falls on land and runs back or melts back into the ocean. It's the same water doing the same thing over and over again. No new water is added and no old water disappears.

Not only is the earth's surface mostly covered with water, our foods are made mostly of water, too. Watermelons, cucumbers, and strawberries are almost all water. So are you, your family, your pets, your friends, and all of us—we are all mostly water. For instance, there is more water in the body of a dog than all the rest of what is in its body.

All living things on Earth must have water. Even desert plants and animals, which get along with little water, must have at least that little bit to survive. People use a lot of water for their personal needs, industry uses 5 times that much, and farmers use even more for irrigating their land. That's why scientists worry about the pollution of our fresh water and are always working on ways to avoid it.

Much of our water, once pure and fresh, is now dirty and smelly. It is polluted and is unhealthy for fish, plants, animals, and people. For years some factories have dumped chemical wastes into the waters. Many home sinks and toilets empty into sewers. This waste, mostly untreated, goes into rivers and lakes. Careless campers and careless boaters add to the mess. But there's hope. Scientists can now make polluted water clean again. Great harm has been done to the waters of the earth, but pollution can be stopped if science, industry, government, and all people work together.

The OCEANS are linked to one another and form one continuous body of water which covers more than 70% of Earth's surface. Because water changes temperature more slowly than air and land, the oceans help control Earth's climate by keeping the air above them at a fairly steady temperature.

Today ocean water is salty, but when the oceans were first formed they were fresh and unsalty. Over millions of years, rain dissolved salts and other minerals in rock and soil, and rivers carried these tiny amounts of dissolved material into the oceans. Now there are 3½ pounds of salt and other minerals in each 100 pounds of ocean water. Drinking salty sea water only makes you more thirsty; people have died of thirst surrounded by ocean water.

By about 1850, the oceans had been well travelled and their boundaries mapped by explorers. But it has only been in the last 100 years that ocean scientists (oceanographers) have begun to find out what lies beneath the surface of the water and what the ocean floor is like. They have found that if the water were removed from the oceans, the landscape would be very much like what we know on land, with plains, plateaus, and mountain chains. They have found that the oceans are full of an astounding variety of living things, from tiny drifting plankton (microscopic animal and plant life) to the huge 100-foot-long blue whale and the giant kelp, a kind of seaweed which can grow to 200 feet long. There are 20,000 kinds of fish (people don't eat very many kinds). There are mammals who come up to the ocean's surface to breathe—seals, whales, dolphins, and porpoises. There are sea turtles and 50 kinds of sea snakes. There are lobsters, oysters, and clams.

The many oceans are joined to form one continuous ocean.

Deep down the ocean is actually very dark.
No light penetrates deeper than about 600 feet.

Smaller fish are food for larger fish.

Scientists who go down into the ocean to study it must bring oxygen with them. They must also protect themselves from the water pressure if they go down very deep. About 3,000 feet down, the pressure is 100 times as much as it is at the surface. Divers' suits and underwater research vehicles protect the ocean scientists.

This is an underwater house that rests on the ocean bottom. Scientists can live in it for months and swim out to do their studies.

fantastic deep-water fish

The bottom of the ocean was once thought to be flat. Now, thanks to exploration and mapping with scientific instruments, we know there are deep canyons, and mountains higher and more rugged than those on land.

A SEASHORE is a place where the ocean and land meet. There are different kinds of seashores. Some are solid rock cliffs with no beach at all. Some are beaches made of large stones, gravel, or pebbles. Some are stretches of soft, nice-to-lie-on sand; those are the beaches people like.

On some beaches the sand is coarse, on others it is very fine. All sand is ground-up rock, worn off the shore or carried into the sea by rivers and then washed up on the beach by waves. The kind of rock determines the color of the sand.

Beaches change every day; they are narrower at high tide and wider at low tide. They become even narrower in the winter months when it is stormy and big waves wash some of the sand out to sea. Sand washed offshore during winter storms piles up in banks under the water. The following summer, the sand from the banks is washed ashore again.

People like to go to the seashore to relax in the sun, swim in the salt water, fish in the surf, collect shells, build sand castles, or hike. If you go to the beach, be careful not to get too much sun; the reflection of the sun's rays off the water increases your chance of getting a sunburn. Also, be sure to swim

Winds pile up the sand at the back of the beach into dunes. Beach grass holds the dunes in place.

These are some of the shells that you can find on many beaches:

clam *scallop* *moon snail* *mussel* *periwinkle* *oyster*

where the lifeguards can see you or where a good swimmer is watching you.

Winds are what make the ocean waves wave. (If you blow gently into a bowl of soup, you can make your own miniature waves in pretty much the same way that the wind makes ocean waves.) Once a wave is formed by the wind, it may move thousands of miles across the ocean until it nears a shore. When it reaches shallow water, the crest tips forward and it tumbles into a foamy breaker. During ocean storms, waves can be very destructive when they crash onto a beach.

There are other beaches made of . . .

green sand . . .   black sand . . .   pink sand . . .

pebbles . . .   and some made of large cobblestones.

See the ship disappear over the horizon.
This happens because the earth is round.

This type of beach slopes down gently to the water. Low and high tides change the width of the beach twice in 24 hours.

Most waves are caused by the wind. They break when water depth is too shallow in relation to their height.

RIVERS often begin on a mountain, hill, or perhaps at the melting edge of a glacier. A river starts as a little stream, which flows toward lower levels and is joined by other streams which make it wider and deeper. Underground water may flow into the stream from springs. When the stream gets big enough, we begin to call it a river. Sometimes the river's flow is interrupted by a lake, but usually it continues until it finally flows into the ocean.

As the river flows down through the same channel or riverbed year after year, it carries sand and pebbles along with it. They act as digging tools, banging and chipping at the banks and bottom of the river. This wearing away of the land is slow, but if it lasts long enough a valley forms. Even the tremendous Grand Canyon was formed in this way.

There are three kinds of river valleys. A valley is young when a rapidly flowing river cuts sharply

LAKES are large inland bodies of water. The water in a lake is standing. That means it isn't running the way water in a river is. Many fish and plants live in lakes.

Lakes are created in different ways. They may be formed in low places if a blockage occurs in the flow of a stream or river. If a glacier gouges deep places in the land surface, and then the climate gets warmer, the deep places may become filled with water flowing in streams from the melting glacier. Sometimes movements of the earth's crust lower a large area

of land, forming a basin, which then fills with water from streams and rain, making a lake. A crater lake forms when an extinct volcano's crater fills with water.

All new lakes are filled with fresh water, but if a lake is drying up or if there isn't a stream through which the water can leave, the lake may gradually become a salt lake. This happens because tiny amounts of dissolved salt from the soil keep seeping in with no way to get out. The Great Salt Lake in Utah is six times as salty as the oceans.

down into it. A valley is called mature when the river cuts sideways into the valley walls and wanders back and forth. An old valley is wide and flat, with a slow-moving river looping through it.

People used rivers for travel and for carrying supplies long before roads were thought of. Today, although people travel by cars, planes, and trains, rivers still carry goods and materials of all kinds, as well as pleasure boats.

BAYS are bodies of water partly enclosed by land, with an opening to the ocean or other large body of water such as a lake or a gulf. A gulf is like a bay, but bigger.

Some bays are much safer places to be than the open sea in bad weather, and boats will usually head into them when the ocean gets rough.

Some bays have amazing tides. The Bay of Fundy between New Brunswick and Nova Scotia in Canada is famous for its tides. At high tide the water in the bay is 50 feet higher than it is at low tide!

PONDS are small, shallow lakes. Usually a pond is formed the way a lake is. Sometimes a pond is all that remains of a disappearing lake.

Ponds are usually no deeper than 15 feet, so the sun's heat gets to the bottom. In summer, sunfish and minnows, which like warm water, live in ponds. Many plants grow there. If there were no plants there would be no fish, for plants give off the oxygen that fish need to stay alive. Ducks nest near ponds and paddle on them, searching for fish. Frogs and toads live in and out of pond water.

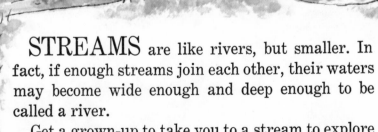

STREAMS are like rivers, but smaller. In fact, if enough streams join each other, their waters may become wide enough and deep enough to be called a river.

Get a grown-up to take you to a stream to explore to see water snakes, lizards, tadpoles, and frogs. Streams are popular with fishermen, who wade in with hip boots and cast their lines for trout.

Will overcrowding on Earth force people to live on satellites out in space?

Will we, at long last, make contact with someone, anyone, out there on another planet?

What new hills and mountains will rise up to replace our present ones?

Will all the continents, drifting apart, finally meet and again form one giant continent in a brand-new arrangement?

EURASIA

ANTARCTICA

NORTH AMERICA

AUSTRALIA

AFRICA

INDIA

And will overcrowding also force people to live in ocean-bottom communities?

SCHOOL

SEA BOTTOM HOMES WELCOME

Yes, THE FUTURE lies ahead. In that future, what changes will take place here on Earth? Will there be new and different mountains? Will the continents continue their slow drifting? Will people live underwater or in outer space? What will happen when our supplies of oil, coal, and gas are gone?

No one knows for sure what will happen in the future, but we can guess that certain things that are already happening will keep on happening. New mountains will keep on rising and old ones wearing away, and our continents will keep up their snail-paced drift of a few inches a year.

If Earth gets too crowded or polluted, should we settle on another planet? Which one should we choose? The other planets in our solar system lack the right air, water, and gravity for our kind of life. What about a planet outside our solar system?

Planets are satellites of stars, and the nearest star is 6,000,000,000,000 (six trillion) miles away from us. If this star had an empty Earth-like planet, should we set out for it?

If we did rocket off to that planet, say at 50,000 miles an hour (twice moon-flight speed) our rocket would arrive 13,000 years from now, so we had better forget that idea! If people ever do live off the earth, it will probably be on nearby space stations.

In the future, the foolish polluting of the waters and air that keep us alive must end. Also we must develop other energy sources such as the sun, wind, and Earth's interior heat, to replace the oil, gas, and coal that we are using up so recklessly.

And in that future, will there be another book to explain some of the new things about our universe that we will have learned by then?

You go into the ocean for a dip,
but a wave knocks you over.
What makes the waves wave?
(See page 65)

You live in the Northern Hemisphere and are
studying the stars. You know that their
position in the sky changes throughout the year.
Yet one of the stars seems almost stationary.
Why? Which star is it?
(See page 8)

You are out for a drive on the moon.
You come to the highest mountains.
How high are they compared to those on Earth?
(See page 14)